T0215312

ROUTLEDGE LIBRARY EDITIONS:
URBANIZATION

Volume 7

THE GEOGRAPHY OF URBAN-RURAL INTERACTION IN DEVELOPING COUNTRIES

THE GEOGRAPHY OF URBAN-RURAL INTERACTION IN DEVELOPING COUNTRIES

Essays for Alan B. Mountjoy

Edited by
ROBERT B. POTTER AND TIM UNWIN

Routledge
Taylor & Francis Group

LONDON AND NEW YORK

First published in 1989 by Routledge

This edition first published in 2018
by Routledge
2 Park Square, Milton Park, Abingdon, Oxon OX14 4RN

and by Routledge
711 Third Avenue, New York, NY 10017

Routledge is an imprint of the Taylor & Francis Group, an informa business

© 1989 Robert Potter and Tim Unwin

British Library Cataloguing in Publication Data
A catalogue record for this book is available from the British Library

ISBN: 978-0-8153-8014-6 (Set)
ISBN: 978-1-351-21390-5 (Set) (ebk)
ISBN: 978-0-8153-7960-7 (Volume 7) (hbk)
ISBN: 978-0-8153-7962-1 (Volume 7) (pbk)
ISBN: 978-1-351-21538-1 (Volume 7) (ebk)

Publisher's Note
The publisher has gone to great lengths to ensure the quality of this reprint but points out that some imperfections in the original copies may be apparent.

Disclaimer
The publisher has made every effort to trace copyright holders and would welcome correspondence from those they have been unable to trace.

THE GEOGRAPHY OF URBAN-RURAL INTERACTION IN DEVELOPING COUNTRIES

ESSAYS FOR ALAN B. MOUNTJOY

Edited by
ROBERT B. POTTER
AND TIM UNWIN

ROUTLEDGE
London and New York

First published 1989
by Routledge
11 New Fetter Lane, London EC4P 4EE
29 West 35th Street, New York, NY 10001

© 1989 Robert Potter and Timothy Unwin

Printed and bound in Great Britain by
Mackays of Chatham PLC, Chatham, Kent

British Library Cataloguing in Publication Data

The Geography of urban-rural interaction
in developing countries.
1. Developing countries. Economic
conditions. Geographical aspects
I. Potter, Robert B. II. Unwin, Timothy,
1955-
440.9172'4

ISBN 0-415-00444-6

Library of Congress Cataloging-in-Publication Data

ISBN 0-415-00444-6

CONTENTS

Contents

PREFACE

In 1985, Alan Mountjoy retired after a long career in the Department of Geography at Bedford College (now Royal Holloway and Bedford New College), University of London. During the time he was teaching, researching and writing, Alan Mountjoy did a great deal to popularise the study of the social and economic geography of developing countries, in schools, as well as in universities and polytechnics. This *festschrift* is being produced in recognition of his contribution, and to mark the occasion of Alan's formal retirement from academic life.

We were anxious, though, to prepare a *festschrift* with a difference. Rather than being a collection of quite disparate research papers, we aimed to bring together a series of chapters written around a central theme of current importance in the geographical study of developing countries. We were also keen that the contributors should all have been associated, at one time or another, with Alan and his teaching and research within the orbits of Bedford College in particular, and the University of London in general. In the event, we were indeed pleased that past colleagues, former students and friends of Alan were so keen to contribute to this volume.

As well as being of some currency, it was also felt appropriate that the theme selected for the book should reflect Alan Mountjoy's concerns with the geography of the Third World. Accordingly, it was decided to provide an introductory text on the topic of urban-rural interaction in developing countries. Within the general field of the geographical study of development, this is an important, although as yet, surprisingly neglected area of teaching and research. The rationale for the volume is to exemplify, both historically and in contemporary spatial terms, the nature and salience of the interactions which occur between so called 'urban' and 'rural' areas within Third World national territories. All too often, in past teaching, research and policy-making, urban and rural have been dealt with as if they denote entirely dichotomous categories, whereas in reality, they are but two sides of the same coin. The differences which characterise the rural and urban tracts of developing countries, in structural, demographic and behavioural terms, are the reflection of a unified set of dynamic processes, which are commonly referred to as 'development' or 'change'. Teaching, research and urban policy all need to reflect this essentially unified perspective. The principal forms of interaction are well-known and easily identified, but the relations existing between them make it difficult to discern the precise outcome of the dynamic processes associated with them. The main types of interaction involved are the movement of people, goods, money, capital, new technology, information and ideas.

Although the rise of structuralist and political economy approaches to development studies has witnessed a renewed interest in rural-urban relations, as yet this is poorly articulated in the empirical research literature. At the same time, there is an increasing awareness of the need for more comparative, cross-cultural and cross-national research in the field of development geography. It is intended that the present chapters, covering as they do, the empirical realities of rural-urban interaction in a variety of African, Middle Eastern, Asian and Caribbean countries, will help to provide much-needed fresh material making-up for these earlier shortcomings.

In preparing the book we have received help, services and encouragement from a number of people. David Bowen, Head of the Department of Geography at Royal Holloway and Bedford New College was supportive from the start. Kathy Roberts produced successive versions of the manuscript with great efficiency, along with a good measure of humour. Roy Davies of the Audio-Visual Centre of the

Preface

College processed all of the art work. Ron Halfhide and Erica Milwain provided various cartographic services. Finally, as the editors of this volume, we should like to record our personal thanks to Alan Mountjoy for all the encouragement he has given us in the past. We were fortunate to have him as a colleague at a time when, having undertaken postgraduate research in England, we both found ourselves increasingly orienting our teaching and research toward the developing world. The same process of transformation from a primarily First World to a Third World focus during time spent in the Department of Geography at Bedford College is true of at least two other contributors to this volume. Alan's direct influence on such transitions is, of course, hard to estimate, but it is likely to have been substantial. Indeed, in this connection, it is worth recording that one of us took our very first formal course on the Third World with Alan Mountjoy as an undergraduate in the Department of Geography at what was to become Royal Holloway and Bedford New College.

Robert B. Potter and Tim Unwin

Egham, Surrey

Alan B. Mountjoy, M.C., M.A. (Reading)
Emeritus Reader in Geography, University of London

ALAN B. MOUNTJOY: AN APPRECIATION

DAVID HILLING

In the summer of 1985, it fell to me as Acting Head of the Department of Geography at the newly created Royal Holloway and Bedford New College, to bid farewell to Alan B. Mountjoy on the occasion of his retirement after thirty-nine years of very devoted service to his subject, his Department, and the University of London. It gave me great pleasure to be invited by the editors of this volume to contribute an introductory appreciation of Alan Mountjoy, which is an expression of the esteem in which he is held by all those who have been fortunate enough to have had academic or social contact with him. For many students, and to his younger colleagues, Alan B. Mountjoy will always be known as 'ABM', and with the greatest respect, I do not intend to depart from this convention.

ABM went up to Reading University in 1937 and immediately became involved in Officer Training Corps activities. Inevitably, on the outbreak of war, he enlisted for military service, but with the war still in its 'phoney' phase, he was able to return to Reading to complete his degree, which he did in 1940, obtaining a First.

Philosophising and conceptualisation have found little place in ABM's work and his research and writing are more the result of instinctive reaction to the events and experiences in which he was involved. The word 'paradigm' was not for him! While at Reading as a student ABM had come under the influence of A. Austin Miller, the physical geographer with special interests in climatology, and George H.T. Kimble, the historical geographer who later, while Chairman of the Geography Department at Indiana University, was to edit the two-volume Twentieth Century Fund sponsored study of Tropical Africa. ABM always

1

Alan B. Mountjoy: an Appreciation

claims that his grounding in physical and historical geography (in his words, "development geography in another guise") had a profound influence on his later approach. Without being deterministic, he insists that activities in the human sphere cannot be divorced from the physical stage upon which they are enacted. For him, a main role of the geographer is to ensure that those in other disciplines connected with development keep their feet firmly on the ground.

This same concern with both the physical and human aspects of the subject will certainly be remembered by generations of students who did fieldwork with him in Pembrokeshire or Cornwall and were required to 'do' the physical geography, even if that meant military-style confrontations with local farmers. The same basic approach is also clearly demonstrated in his latest book, *Africa - Geography and Development* (with David Hilling, Hutchinson, 1988) in which the physical geography has a prominence which many may think excessive. ABM will doubtless have a wry smile at the thought of the many geography departments now trying to provide for an 'integrated approach'.

On graduating in 1940, ABM went straight into the army and, as a Regimental Signals Officer, was sent on an instructors' course at Catterick where he 'graduated' second out of 120. This was to be his only formal instruction in how to instruct - but was to be that much more than most of his later University colleagues received. In the Autumn of 1943, he was posted to Oran, but with the North Africa campaign concluded, he was able to spend some time travelling in Algeria and Tunisia before joining the 78th 'Battleaxe' Division, first as a mortar officer and later, after Monte Casino, as a Company Commander. The Italian campaign took him north to the Po valley and after the cessation of hostilities, he spent periods of time in both Austria and Germany.

However, in the Autumn of 1944, he was posted to Egypt for three months. As a geographer, he was immediately impressed by the apparent contradiction between the potential prosperity and actual poverty, and in particular, by the fact that children everywhere appeared in large numbers. This stimulated interest in North Africa in general, Egypt in particular, and the demographic aspects of the development

2

Alan B. Mountjoy: an Appreciation

problem which were to become the cornerstones of his work in the years after the war.

Leaving the army in 1946, having been decorated with the Military Cross, ABM was appointed to the lecturing staff at Bedford College, University of London, where he was to remain for his whole teaching life. He started work on an M.A. by thesis for Reading University, on 'Factors in the Economic Development of Egypt' and this was completed in 1949. In the same year, a paper on the 1947 population of Egypt (*Geography* 68, p.13-21) was the first of many that would be concerned with various aspects of the population problems of Egypt and developing countries in general, and his 1952 paper on 'Problems of industrialisation: an Egyptian example' (*Souvenir Volume, Indian Geographical Society*, p.14-25) pointed to another subject area in which he was to become authoritative over time.

During his early years at Bedford College, ABM's teaching had been concerned primarily with economic geography, but in 1956, he introduced an Africa regional geography course, which increasingly reflected his special interest in population and development issues. In 1961, an invitation by Professor S.W. Wooldridge, geography editor of the Hutchinson University Library series, to produce an economic geography text gave ABM the opportunity to write a more specialised book which reflected his interest in industrialisation in developing countries. In the early 1960s, few geographers had made any mark in development studies, and on publication in 1963, *Industrialisation and Under-developed Countries* was in many respects a pioneering work. While it was clear that industrialisation had a critical role in the development process, it was not the universal panacea claimed by the wisdom of the time. ABM's well-rounded discussion, placing industrialisation in its broader development context, did much to provide a more balanced view and, in particular, pointed to the relevance of a geographical approach in development studies. The book was to make ABM well-known world-wide and it says much for the soundness of his work that twenty-five years and five editions later the book is still without competitors.

ABM's interest in Africa continued to be reflected in articles on North Africa and in the publication with C. Embleton of *Africa: a Geographical Study* (Hutchinson, 1966). This book also went to three editions and like

Alan B. Mountjoy: an Appreciation

Industrialisation and Developing Countries, merited American versions. Kindled by his wartime experiences, ABM's interest in Italy was maintained by return visits in most years and reflected in numerous journal articles on aspects of the economic geography of the country. In Italy's 'South', ABM saw a mirror of conditions and processes in developing countries and a more convenient laboratory in which to pursue his interests. This led to his study *The Mezzogiorno* (Oxford, 1973).

In the mid-1960s, the unitary University of London geography programme was replaced by College-based courses which allowed greater flexibility and the development of specialist interests based on staff expertise. ABM became the senior member of an enlarged human geography group at Bedford College and he introduced a new course on The Economic and Social Geography of Developing Countries. He was always eager to involve as many colleagues as possible, with physical and human interests, in this course and was concerned that teaching was based on first-hand experience in developing countries. The course attracted large numbers of undergraduates and in modified form, is still one of the Department's most popular. The Africa and West Mediterranean courses became ever more development oriented.

Between 1967 and 1970, ABM acted as the External Examiner at the Department of Geography in the University of Khartoum and he visited Sudan each year. It is particularly gratifying that this volume includes an essay by Professor El-Bushra, now of the University of Riyadh, Saudi Arabia, the first of a succession of Sudanese students who undertook postgraduate research in the Bedford College department under ABM's supervision. In the late 1970s, ABM became the Department's Postgraduate Tutor and large numbers of students were attracted for M.Phil./Ph.D. research and also M.A./M.Sc. taught programmes with an emphasis on development studies. With sometimes as many as seven countries represented, the postgraduate seminars could be a forum for the lively debate of development issues. Certainly, ABM took great pride in the fact that the Bedford College Geography Department on several occasions topped the geography departments listed in the Institute of Development Studies Research Directory.

Alan B. Mountjoy: an Appreciation

ABM would take great satisfaction and find no academic slight in being described as a 'populariser' of his subject. Most of his articles appeared in the more widely read and possibly less erudite journals and he was a great supporter of Derek Weber's efforts to make the *Geographical Magazine* more academic in appeal. ABM was a frequent contributor to the *Geographical Magazine* and a firm believer in getting reasonably academic papers into a popular format with wide circulation. In 1970, he edited a series of articles in the magazine, many written by Bedford colleagues, on "Developing the Under-Developed Countries" and these eventually appeared in book form, *The Third World: Problems and Perspectives* (Macmillan, 1978) published in both London and New York. The same concern with getting 'good' geography to the widest possible readership was manifest from the late 1950s, when he assisted with editing geographical material for the *Encyclopaedia Britannica* and more recently, when he acted as Consultative Editor to the Reader's Digest *Guide to Places of the World* (1987).

In 1980, ABM was invited to contribute regularly to *Third World Quarterly* and saw this as a reflection of the high standing of the Department and also as an opportunity to provide an overtly geographical input to the wider development discussion. The outcome was a series of world maps, researched and prepared in the Department, on various aspects of development, each map having an accompanying explanatory text and serving to highlight global contrasts and national individualities. This he saw as part of the geographers' role - to assist in keeping other peoples' feet on the ground.

An extended visit to Papua New Guinea in the early 1980s allowed ABM an opportunity to expand his growing interest in the operational patterns and impact of multi-national corporations on the development process - successive editions of *Industrialisation and Developing Countries* having devoted more attention to this topic. In 1984, this work was published as his last major paper, 'Core-periphery, Government and multinational: a Papua New Guinea example'.

In his earlier years at Bedford College, ABM was actively involved in the Institute of British Geographers and when the annual conference was held at the College in 1953, he acted as the local organising secretary. He was also involved

Alan B. Mountjoy: an Appreciation

in some of the earliest discussions that were eventually to lead to the creation of the IBG's Developing Areas Research Group. In 1964, when the International Geographical Union held its Congress in London, ABM acted jointly with Eric Rawstron as secretary of the Economic Geography section. Although there was no development studies component as such for the London meeting, ABM used the occasion to include papers that were in effect just that. He was also a contributor to the *Congress Excursions Guide*.

For many years, ABM was in great demand to lecture at Geographical Association meetings and he identified closely with the educational function of that organisation. At such meetings he found 'question time' the most refreshing part and it was greatly to his regret that increasing deafness forced him to take less and less part in conferences and meetings of all kinds. However, he did maintain his examining commitments, and was for many years until his retirement, the Chairman of the University of London External Board of Examiners in Geography.

ABM has no sympathy with what he sees as an excessive emphasis on theory and modelling in development studies and feels strongly that the failure of so many development policies is a direct conseqeunce of their failure to relate to real life conditions and a general ignorance on the part of development theorists of the human and environmental factors that are involved. In his research, writing and teaching, ABM was always concerned with actual examples and real problems. His students will certainly remember the anecdotal element which illuminated his teaching. His approach to development studies was firmly based on an appreciation of the individual place and conditions, and the need to seek particular rather than generalised solutions to development problems. Perhaps development progress might have been more substantial, if more decision-makers paid attention to his dictum that "Every nation runs an individual course in the development cross country".

As a teacher, supervisor, colleague and friend, ABM was unfailingly sympathetic and supportive and those who have worked with him have come to value highly his knowledge and understanding and the dryness which characterises his considerable sense of humour. He typified everything that was best in being described as an officer and a gentleman

6

and earned a universal respect which we all would wish for but few achieve.

That we still attract to Royal Holloway and Bedford New College large numbers of students interested in development studies is eloquent testimony to the firmness of the foundations established by ABM, and for this we are fully appreciative. ABM also provided us with the salutary lesson, yet to be fully appreciated by our political and possibly administrative masters, that a world-wide reputation of the highest order can be based on sound achievement with minimal financial research support and with a modest rather than high profile.

ALAN B. MOUNTJOY: PUBLICATIONS

1949 'A note on the 1947 population of Egypt', *Geography* 68, 13-21
1950 *Chambers Encyclopaedia*, Bulgaria
1952 'Problems of industrialisation: an Egyptian example', *Souvenir Volume*, Indian Geographical Society, 14-25
1952 'In the Wicklow Mountains', *Canadian Geographical Journal* 44, 242-51
1952 'The development of industry in Egypt', *Economic Geography* 28, 212-48
1953 'Egypt's population problem', *Transactions of the Institute of British Geographers* 18, 121-35
1953 *London and South-east England*, London
1955 'Land and the Egyptian peasant', *Geography* 40, 194-96
1956 *Victoria Encyclopaedia*, African and Asian entries
1957 'Vegetable oils and oilseeds: aspects of production and use', *Geography* 42, 37-49
1957 'The Suez Canal, 1935-55', *Geography* 42, 186-90
1957 *The Midlands*, London
1958 *Farming in the Paris Basin*, Filmstrip with notes, pp.20, London
1958 *The Rhone Valley*, Filmstrip with notes, pp.24, London
1958 'The Suez Canal at mid-century', *Economic Geography* 34, 157-67
1959 *Encyclopaedia Britannica*, Algeria
1959 'Milk pipelines and mountain economies', *Geography* 44, 124-26
1959 *A Journey Through South West France*, Filmstrip with notes, pp.24, London
1960 *Encyclopaedia Britannica*, Egypt
1960 Contributions to Oxford Regional Economic Atlas, *The Middle East and North Africa*, Oxford
1963 *Industrialisation and Under-developed Countries*, London
1963 'Food and agriculture', in *Technology*, London
1964 (with C. Embleton) 'Geomorphology and industry in part of the Middle Thames Valley', *Guide to London Excursions*, 20th International Geographical Congress, 88-92
1965 Contributions to *The Complete Atlas of the British Isles*, Readers' Digest, London
1966 (with C. Embleton) *Africa: a Geographical study*, London
1966 'Planning and industrial developments in Apulia', *Geography* 51, 369-72
1966 *Italy: the changing South*, Filmstrip with notes, pp.39, London
1966 Revision of Laborde's *Western Europe*, London
1967 (with E.M. Rawstron) 'Report of section meetings in Economic Geography', *Congress Proceedings* of 20th International Geographical Congress, 134-43, London
1967 (with C. Embleton) *Africa: A new Geographical Survey*, New York
1967 (with C. Embleton) *Africa: A Geographical Study*, 2nd edition, London
1967 *Industrialisation and Under-developed Countries*, 2nd edition (revised), London
1967 *Industrialisation and Under-developed Countries*, American edition, Chicago
1968 'Million cities: urbanisation and the developing countries', *Geography* 53, 365-74
1969 'Crops for industry', *Mind Alive* 64, 1773-76
1969 'Natural fibres', *Mind Alive* 65, 1805-08
1970 Editor and contributor, *Developing the Under-developed Countries*, London
1970 'Growing cities', *Geographical Magazine* 42, 357
1970 'Industrial development and planning in Eastern Sicily', *Geography* 55, 441-44

Alan B. Mountjoy: Publications

1970 'Pressures and progress in southern Italy', *Geographical Magazine* 42,
1970 (with C. Embleton) *Africa: a Geographical Study*, 3rd edition (revised), London
1971 *Industrialisation and Under-developed Countries*, Portuguese edition (for Brazil)
1971 'Re-opening the Suez Canal', *Geographical Magazine* 43, 649-54
1971 *Industrialisation and Under-developed Countries*, 3rd edition (revised), London
1971 'Egypt', etc. Contributions to Elsevier International Encyclopaedia of Geography, *Geography and Man*, London
1971 'New life for the Suez Canal?', *Contemporary Review* 219, 263-68
1971 'Revolt in Reggio di Calabria', *Geographical Magazine* 43, 516
1972 Contributions to *The Atlas of the Earth*, London
1972 'Egypt cultivates her deserts', *Geographical Magazine* 44, 241-50
1972 'Egypt: population and resources', in: Clarke and Fisher (eds.) *Population of the Middle East and North Africa*, London, 291-314
1972 'Water policy unites the Sudan', *Geographical Magazine* 44, 705-11
1973 *The Mezzogiorno*, Oxford
1973 'The Third World', *Geographical Magazine* 45, 423-32
1974 'Urban explosion in the Third World', *Geographical Magazine* 46, 208-16
1975 *Industrialisation and Developing Countries*, 4th edition (revised), London
1976 'A note on Italy's energy problems', *Geographical Magazine* 48, 433
1976 'Urbanisation, the squatter and development in the Third World', *Tijdschrift voor Economiche en Sociale Geografie* 67, 130-37, reprinted in: Bourne, L.S. and Simmons, J.W. (eds.) *Systems of Cities*, New York, 1978
1976 'Kryzys urbanizacjii w krajach trzeciego swiata' (Urban crisis in the Third World), *Przeglad informacjii o Afryce* 41, 3-20
1977 'The Snowy Mountains hydro-electric project', *Geographical Magazine* 49, 424-29
1977 'Industry in developing countries', *Planet Earth* 84, 1661-65
1978 Editor and contributor, *The Third World: problems and perspectives*, London
1979 'Taiwan alone', *Geographical Magazine* 51, 464-70
1979 Editor and contributor, *The Third World: problems and perspectives*, New York
1980 'Housing and new towns in Hong Kong', *Geography* 65, 53-57
1980 'World population increases', *Third World Quarterly* (map with explanatory text)
1980 'World food supply: protein intake', *Third World Quarterly* (map with explanatory text)
1980 'Squatters in Hong Kong', *Geographical Magazine* 53, 119-25
1980 'World health services: hospital facilities', *Third World Quarterly* (map with explanatory text)
1980 'Worlds without end', *Third World Quarterly* 2, 753-57
1980 (with A. Homoudi) 'Egypt', 'Sudan', *Philip's Illustrated Atlas*, London
1981 'International aid: donations and receipts', *Third World Quarterly* (map with explanatory text)
1981 'World trade in crude oil', *Third World Quarterly* (map with explanatory text)
1981 'World levels of literacy', *Third World Quarterly* (map with explanatory text)
1981 'World energy consumption', *Third World Quarterly* (map with explanatory text)
1982 *The Mezzogiorno*, 2nd edition (revised), Oxford

9

Alan B. Mountjoy: Publications

1982 'World medical services: numbers of doctors', *Third World Quarterly* (map with explanatory text)

1982 'Third World indebtedness', *Third World Quarterly* (map with explanatory text)

1982 *Industrialisation and Developing Countries*, 5th edition (rewritten and enlarged), London

1983 'Third World military spending', *Third World Quarterly* (map with explanatory text)

1984 'Core-periphery, Government and multinational: a Papua New Guinea example', *Geography* 69, 234-43

1985 'A decade alone for PNG', *Geographical Magazine* 57, 462-63

1986 'The progress of world urbanization', *Geography* 71, 246-48

1987 Consultant editor, *Guide to Places of the World*, Readers' Digest, London

1988 (with D. Hilling) *Africa: Geography and Development*, London and Totowa

1

URBAN-RURAL INTERACTION IN DEVELOPING COUNTRIES: A THEORETICAL PERSPECTIVE

TIM UNWIN

Despite a long tradition of geographical research into rural-urban migration, there has until recently been a dearth of material published on wider *interactions* and *linkages* between urban and rural areas. Instead, the bulk of research has been devoted to the analysis of urban and rural 'development' as separate issues (Harriss, 1982; Chambers, 1983; Roberts, 1978; Gilbert and Gugler, 1982; Potter, 1985). Since 1980, though, a growing awareness of the importance of urban-rural relationships, and a dissatisfaction with urban-based centralised models of 'development' has led to a considerable theoretical reappraisal of such models and issues (Dixon, 1987). As O'Connor (1983, p.274) has noted in the African context, "these rural-urban linkages are matters on which very few precise data are available". Consequently, one of the purposes of this *festschrift* for Alan Mountjoy is to provide specific case studies exemplifying the nature of urban-rural interactions in different parts of the world, against which the growing body of theoretical literature can be appraised. The introduction is designed to provide a brief overview of some of the more important theoretical arguments which have recently addressed the subject of urban-rural interaction, and it also places the individual essays which comprise the book within the context of this broader literature.

The concentration of much of the 'development' literature on urban and rural issues as *separate* concerns has had unfortunate consequences both for the world's poor and also for our understanding of the processes involved in social and economic change. By focusing on separate urban and rural themes, attention has been drawn away from the connections *between* these two loci of change. Increasingly it is now being argued that urban and rural change should be seen, not as processes in themselves, but rather as the *products* of deeper structural transformations in society. This reorientation of attention has enabled different kinds of research agenda to be formulated. By concentrating on the linkages and flows between towns and countryside a more comprehensive grasp of the processes of social and economic change affecting the world's poor may be aehieved.

Such an argument has been well expressed by Harvey (1985, p.14-15) in his study of the urbanisation of capital when he suggested that focusing on the urban-rural "dichotomy may be useful in seeking to understand social formations that arise in the transition to capitalism - such as those in which we find an urban industrial sector opposed to a rural peasant sector which is only formally subsumed within a system of commodity production and exchange. But in a purely capitalist mode of production - in which industrial and agricultural workers are all under the real domination of capital - this form of expression of the division of labour loses much of its particular significance". Urban and rural change therefore need to be seen as parts of an overall social formation, and one way in which this can be achieved is through an analysis of the interactions between these two foci of change. It is with the nature and significance of such interactions that this book is concerned.

In so doing, it builds on a small, but growing body of literature specifically concerned with rural-urban interaction which has emerged in the last few years. One of the first attempts to define this interaction was undertaken by Preston (1975), who identified five main categories of interaction: movement of people, movement of goods, movement of capital, social transactions, and administrative and service provision. More recently in the British context Gould (1982, p.334) has recorded how "rural-urban interaction was the theme of the first of a series of four workshops on the Third World to be sponsored during 1982 by the Human Geography Committee of the SSRC" (Social Science Research Council). This workshop highlighted three principal aspects of interest in the subject: population mobility, resource transfer, and social interaction. These interests were further developed in 1984/85 by the establishment of a Rural-Urban Interaction

Network by the Developing Areas Research Group of the Institute of British Geographers. This was designed to elicit research proposals concerned with the issue of rural-urban interaction in the Third World. Within this context, Gould (1985, p.1) noted that rural-urban interaction could be considered "as the two-way flow of people, goods, money, technology, information and ideas between rural and urban areas, themselves seen as metaphors for a wider range of spatial categories at various scales", and he further suggested that "these flows are not only symptoms of the 'development process' but are themselves active features in the transformation of rural and urban places" (Gould, 1985, p.1). Such ideas have in turn been followed up in Dixon's (1987) edited volume of case studies of rural-urban interaction in various parts of the so-called developing world.

Underlying this shift in attention to a specific concern with linkages and flows between the countryside and towns has been a theoretical reappraisal of certain urban-based models and ideas which have been used in planning policies by governments throughout the world over the past three decades. Three particular recent critiques, those by Lipton, Rondinelli, and Stöhr and Taylor, each addressing the same central problem but with different individual aims and emphases, warrant particular consideration, and are addressed in this chapter. However, before this is undertaken, it is essential to clarify several of the key concepts with which they are concerned.

Growth Poles, Top-Down Development and Parasitic Cities

Three basic and interrelated ideas have dominated much of the literature on urban-rural links in development planning since the late 1950s: the *growth pole* concept, the distinction between *top-down* and *bottom-up* development, and the conceptualisation of cities as being either *parasitic* or *generative*. Each of these requires some elaboration. Rondinelli (1985, p.3) has succinctly captured the main aspects of the first of these in the following terms: "The growth pole concept of spatial development suggests that by investing heavily in capital-intensive industries in the largest urban centers, governments in developing countries can

13

stimulate economic growth that will spread outward to generate regional development". The growth pole concept can thus be seen to be overtly concerned with *where* 'development' takes place, and explicit within it is the idea that emphasis will be placed on urban initiatives with the expectation that these will in turn generate rural development at both a regional and local scale (Hansen, 1981; Parr, 1973; Thomas, 1972). As Rondinelli (1985, p.4) goes on to observe, the growth pole concept was based upon the principle "that the free operation of market forces would create 'ripple' or 'trickle down' effects that would stimulate economic growth throughout the region. Investment in industry at the growth pole would be the 'engine of development' for agricultural and commercial activities".

Stöhr and Taylor (1981) have argued similarly in relating the growth pole concept to top-down planning. They comment that "development 'from above' has its roots in neoclassical economic theory and its spatial manifestation is the growth centre concept .... The basic hypothesis is that development is driven by external demand and innovation impulses, and that from a few dynamic sectoral or geographical clusters development would, either in a spontaneous or induced way, 'trickle down' to the rest of the system" (Stöhr and Taylor, 1981, p.1). In this context it can also be noted that it is not only the so-called 'free market' capitalist societies that have embarked on such policies. Many centrally planned socialist states have also turned towards top-down, centralised planning strategies. The implications of such development policies for urban and rural areas are that investment should be primarily urban and industrial in nature, and based on large projects using 'high technology'. This, it is argued should in turn lead to rural 'development', which would spread out from the growth pole, and eventually give rise to a convergence of per capita income in different parts of the state.

Experience from Latin America and Africa, where growth pole policies were adopted, has shown that these beneficial 'trickle down' influences have on the whole failed to materialise (Conroy, 1973; Santos, 1975), and that they have instead been replaced by adverse 'backwash' effects which have created increased inequality between cores and peripheries and between urban and rural areas. This

14

distinction between 'trickle down' and 'backwash' effects is closely paralleled by Hoselitz's (1957) classic distinction between 'generative' and 'parasitic' cities: 'generative' cities are those which are responsible for beneficial influences, whereas 'parasitic' cities are those that give rise to adverse effects in their surrounding rural regions. Within this framework the critical features that need identifying are thus the conditions and linkages that make some cities generative and others parasitic.

The failure of *top-down* planning, based on the growth pole concept, has itself given rise to alternative *bottom-up* policies which seek to place the emphasis on the provision of basic needs and the overt 'development' of rural areas (Stöhr and Taylor, 1981). This concern with a reorientation of planning emphasis is well expressed in Chambers' (1983) book on *Rural Development: putting the last first*, and Richards' (1985) volume on *Indigenous Agricultural Revolution: ecology and food production in West Africa*. In these works attention is immediately focused on the need for development beginning with the rural poor and with indigenous, local practices and belief systems.

The recent works of Lipton, Rondinelli, and Stöhr and Taylor build on these concepts, and taken together provide a range of theoretical standpoints from which to view urban-rural interaction.

Theories of Urban-Rural Interaction

Lipton and urban bias

One of the most forthright condemnations of the effects of urban-based, top-down development policy in recent years has been Lipton's (1977, 1982) advocacy of the concept of urban bias. This seeks to illustrate that "the most important class conflict in the poor countries of the world today is not between labour and capital. Nor is it between foreign and national interests. It is between the rural classes and the urban classes" (Lipton, 1977, p.13). In developing his argument, Lipton rejects the following views: (i) that there is no clear distinction between urban and rural; (ii) that this binary distinction is too simple; (iii) that the real division lies

15

between capital cities and the rest of their nations; and (iv) that a sectoral industrial-agricultural division is of more significance in explaining the distribution of poverty. In so doing, he seeks to establish grounds upon which it is possible to identify a clear urban bias in the allocation of resources. In essence, Lipton's argument is that the power of urban people is such that they are able to direct a disproportionate share of resources towards their own interests and away from the rural population. Not only does he suggest that this urban bias keeps poor people poor, but he also asserts that "inequalities *within* rural areas also owe much to the urban-biased nature of development policy" (Lipton, 1982, p.68). He argues that this is because of the existence of an alliance between the urban elite and the richer farmers who are able to provide surpluses of food, savings, and 'human capital' to those in the cities. In Lipton's (1982, p.68) words, "The towns get their cheap surpluses, food, exportables, etc., even if not made very efficiently and equitably. The rural better-off get most of what is going by way of rural investment, price support, subsidies, etc., even if not much of these. The rural poor, though efficient, get mainly pious words, though often sincere ones".

Corbridge (1982, p.95) has observed that the symptoms of Lipton's urban bias are "the cheap procurement prices paid by the urban sector for food, and a series of other 'price-twists' adverse to the interests of the rural class; the heavily imbalanced investment strategies favouring the urban/industrial nexus and the resultant rural skill drain; and the basic lack of health care and educational facilities that defines the rural sector". Here then we have a series of factors linking urban and rural areas which can be measured in the context of specific places to evaluate the validity of Lipton's arguments. However, at a more theoretical level, Corbridge (1982) has also identified three broad areas of criticism to which Lipton's "master metaphor" (Harriss, 1982, p.40) of urban bias is susceptible. Firstly, he suggests that it is not possible to speak of undifferentiated urban and rural societies, with the urban population simply gaining at the expense of the rural. Lipton thus seems to pay insufficient heed to the existence of the urban poor and the rural rich. Secondly, Corbridge develops this critique of Lipton's concept of *class*, by pointing out the contradictory position of the rural elite who, with regard to food prices, the provision of transport and educational facilities, become the natural leaders of the 'rural class', and yet who with regard to the provision of agricultural inputs become at the same time part of the 'urban class'. Corbridge's (1982) third criticism of Lipton's approach is its reductionist conception of politics, whereby Lipton is forced to argue that his urban and rural classes each have clearly defined politics and interests, whereas in reality there are rarely such clear-cut urban versus rural political allegiances.

A key feature of Lipton's arguments, namely the existence of specific urban and rural *classes*, is therefore brought into doubt, and consequently it is necessary to question the whole edifice of 'urban bias', which is predicated upon the existence of these classes (Dixon, 1987). At a descriptive and empirical level, though, Lipton has provided a useful account of the relative flows of surpluses between the countryside and the towns. What he has failed to do is satisfactorily to explain *why* these flows occur. Part of the reason for this would seem to lie in his conflation of *people* with *places*. While it is relatively easy to identify places as being either rural or urban, this is a much more difficult exercise as far as people are concerned, and it is the *people*, rather than the places, who are responsible for creating the flows that exist between the rural and urban areas. As O'Connor (1983, p.29) has suggested for Africa, "the debates must therefore be about whether urban dwellers do or do not exploit the rural areas; and the special difficulty in Africa lies in the identification of these two groups".

Again we are brought to the question outlined earlier in this chapter: is it something specifically about urban areas that determines the nature of the flows of resources to people within them, or must we see the relationships between rural and urban areas as being subject to some other underlying structural forces within society?

Rondinelli: secondary cities and the diffusion of urbanisation

One feature which Rondinelli and Lipton have in common is that both of their theoretical positions are claimed to be appropriate regardless of the political and economic structure of the society under consideration. For Lipton it is something about the *urban* population, whether under capitalism or socialism, that determines the pattern of inequitable development; for Rondinelli it is, in essence, the nature of the city size hierarchy that is the critical factor in determining the success or otherwise of development policies. In contrast to advocates of rural-based and bottom-up programmes of development, Rondinelli (1983, p.10) has argued that "rural development goals, no matter how carefully conceived, cannot be achieved in isolation from the cities or entirely through 'bottom-up' strategies". In what is very much a pragmatic stance, he suggests that "Linkages are crucial because the major markets for agricultural surpluses are in urban centres; most agricultural inputs come from organizations in cities; workers seek employment as rising agricultural productivity frees rural labour; and many of the social, health, educational, and other services that satisfy basic human needs in rural areas are distributed from urban centres" (Rondinelli, 1983, p.10). Consequently, Rondinelli (1985) suggests that if governments in developing countries wish to achieve widespread development, in both social and spatial terms, they must develop a geographically dispersed pattern of investment. This, he argues, can be achieved through the creation of "a deconcentrated, articulated, and integrated system of cities" which "provides potential access to markets for people living in any part of the country or region" (Rondinelli, 1983, p.19).

Central to this approach is the argument that neither the diffusion pole nor the parasitic view of *small cities* is appropriate, and instead that "decentralised investment in strategically located settlements can create the minimal conditions that enable rural people to develop their own communities through 'bottom-up' and autonomous processes" (Rondinelli, 1985, p.8).

It is the concentration of Rondinelli's approach on *linkages*, and in particular on linkages both between rural areas and small cities, and on those between smaller and larger cities, that makes it of such interest in any consideration of urban-rural interaction. Rondinelli (1983) sees the development of what he terms 'secondary cities' as providing five main beneficial results. Firstly, he argues that it relieves pressures on the largest cities in terms of problems of housing, transport, pollution, employment and service provision. Secondly, it reduces regional inequalities, since, if it is accepted that the standard of living is higher in urban than in rural areas, then the spread of secondary cities would lead to the spread of the benefits of urbanisation. Thirdly, it is seen as stimulating rural economies through the provision of services, facilities and markets for agricultural products, as well as being able to absorb surplus labour should agricultural production become more labour efficient. Fourthly, it provides increased regionally decentralised administrative capacity, and fifthly, it helps to alleviate poverty in intermediate cities where Rondinelli (1983) argues that the problems of poverty and marginality are often most acute and visible.

In referring to the debate over whether cities are generative or parasitic, Rondinelli (1983) has suggested that the extent to which secondary cities have development influences on their regions depends on the degree to which the following ten factors are applicable: (i) local leaders identify their success with that of their city and region; (ii) local leaders invest in their city; (iii) local leaders are innovative; (iv) local leaders are aggressive; (v) national government supports the internal growth of the city and its region; (vi) the economic activities established within the city are linked to its hinterland through mutually beneficial processes of exchange; (vii) economic activities within the city are linked to each other to generate multiplier effects; (viii) economic activities are organised to generate income for local residents and promote internal demand; (ix) public and private sectors cooperate to promote economic activities that generate widespread participation and distribution of benefits; and (x) the willingness of the city's leaders to promote and encourage social and behavioural changes responsive to new conditions and needs, which in turn are acceptable to the city's residents.

From the above brief summary two points concerning Rondinelli's approach to development planning are clear: it is essentially concerned with the implementation of change through the manipulation of the urban settlement hierarchy, and it is implicitly based on a 'free market', capitalist framework of change. Each of these observations requires some elaboration.

Unlike Lipton, Rondinelli has argued that rural change can best be implemented and encouraged by the provision of social and economic facilities in medium-sized urban settlements. This is closely similar to arguments put forward by Wanmali (1981) who suggests that, in the Indian context, it is the lack of service provision in the smaller towns that has been one of the most important constraints on rural change. Rondinelli's arguments however assume that beneficial rural change will *necessarily* be promoted by the development of such things as marketing facilities within the towns, and yet he fails to provide a convincing argument that the changes instilled will be *mutually beneficial* to people living in both rural and urban areas. Underlying Rondinelli's argument is the idea that there is such a thing as a 'balanced' pattern of urbanisation. Indeed, in suggesting that "many developing countries do not have efficient systems of central places" (Rondinelli, 1983, p.19), he implies that such an "efficient" system can be created. In practice this is seen as being one that closely fits the classical rank-size rule distribution (Zipf, 1949). Rondinelli argues that primate city structures dominate the developing world and that to achieve balanced development it is therefore necessary to implement policies that will enhance the growth of middle ranked cities. For him, "the challenge for international assistance organizations and national governments in the Third World is to find effective and appropriate ways to help local governments and private investors to strengthen the economies and service delivery capacities of secondary cities, both through direct investment and through national policies that have spatial implications" (Rondinelli, 1983, p.196). The problem with this is that there is no real evidence that the classical rank-size distribution does indeed provide the context for 'successful development', however that is defined, nor is there evidence to suggest that such a pattern will necessarily benefit the poor and underprivileged. If anything, the evidence would support a conclusion that in some cases, as in the examples of the Caribbean and the Arabian peninsula discussed in chapters 6 and 9 of this book, it is actually a state of primacy which is associated with rapid *economic* development. The crucial point to note here, though, is that this need not be associated with an improvement in the way of life of the rural poor.

In turning to the 'free market' nature of Rondinelli's approach, it can be seen from the list of factors by which he differentiated between situations in which secondary cities either do or do not have development impacts on their surrounding regions, that most of them depend upon a certain degree of altruism on behalf of the 'urban' population and the leaders of national and local governments. For Rondinelli the success of balanced regional development is thus likely to be ensured when aggressive and innovative local leaders identify themselves with, and invest in, the cities in which they live. In this context it is significant to note that none of the ten features identified by Rondinelli as generating beneficial regional influences is directly concerned with the rural economy, and in a 'free market' system it is difficult to see how those with economic and political power are going to be persuaded to relinquish some of the advantages currently accruing to them for the benefit of the urban and rural poor. Rondinelli (1983, p.12) does, though, specifically claim that "a system of functionally efficient intermediate cities .... can make an important contribution to achieving widespread economic growth and an equitable distribution of its benefits in both capitalist and socialist societies". This implies that it is the urban settlement structure, rather than the underlying mode of production, which is of particular significance in influencing the equitable distribution of resources. However, Rondinelli fails to evaluate the significance of the nature of socialism and capitalism in the 'development process', and throughout his analysis he assumes implicitly that the context within which his secondary cities are to be developed is indeed a capitalist one. Many of his ideas were developed through his involvement in a series of projects in the Philippines, Bolivia and Upper Volta (now Burkina Faso), sponsored by the United States Agency for International Development. These were known as the Urban Functions in Rural Development projects, and the broadly capitalist framework within which they were set will, to an extent, have influenced Rondinelli's wider theoretical arguments. These now need a thorough evaluation in contexts other than just the capitalist one.

Despite these problems, Rondinelli has usefully drawn together a broad basis for the analysis of major linkages in 'spatial development', and this is summarised in Table 1.1. Although the elements included within the linkage types of very different kinds of category, ranging from systems to patterns and interdependencies, this does provide a broad framework within which to view the case studies which make up this book.

Stöhr and Taylor: bottom-up development

If Rondinelli was primarily concerned with the implementation of beneficial change through manipulation of the urban hierarchy, the work of Stöhr and Taylor (1981) represents a complete contrast. They argue that such top-down development policies need to be fundamentally integrated with bottom-up approaches if 'development' is to become more equitable. For them, "development 'from below' considers development to be based primarily on maximum mobilization of each area's natural, human and institutional resources with the primary objective being the satisfaction of the basic needs of the inhabitants of that area" (Stöhr and Taylor, 1981, p.1). It is "oriented directly towards the problems of poverty and must be motivated and initially controlled from the bottom" (Stöhr and Taylor, 1981, p.1). As such it stands in marked contrast to the types of approach advocated by Rondinelli, and is usually rural-centred, small-scale, and based on the use of 'appropriate technology'. Three other features of development from below are identified by Stöhr and Taylor: it is determined from within and is therefore unique to each society; it is egalitarian and self-reliant; and it is communalist. As they argue, "it involves selective growth, distribution, self-reliance, employment creation, and above all it respects human dignity" (Stöhr and Taylor, 1981, p.454). This is therefore a very different type of definition of 'development' from that used by Rondinelli and Lipton.

The shift of balance towards social rather than economic aims is also reflected in Friedmann and Douglas's (1978, p.163) advocacy of agropolitan development where the "primary objective is no longer economic growth but social development with focus on specific human needs". For them

Table 1.1: Rondinelli's classification of major linkages in spatial development

Linkage Type	Elements
Physical linkages	Road networks, River and water transport networks, Railroad networks, Ecological interdependencies
Economic linkages	Market patterns, Raw materials and intermediate goods flows, Capital flows, Production linkages - backward, forward and lateral, Consumption and shopping patterns, Income flows, Sectoral and interregional commodity flows, "Cross linkages"
Population movement linkages	Migration – temporary and permanent, Journey to work
Technological linkages	Technology interdependencies, Irrigation systems, Telecommunications systems
Social interaction linkages	Visiting patterns, Kinship patterns, Rites, rituals, and religious activies, Social group interaction
Service delivery linkages	Energy flows and networks, Credit and financial networks, Education, training and extension linkages, Health service delivery systems, Professional, commercial and technical service patterns, Transport service systems
Political, administrative and organisational linkages	Structural relationships, Government budgetary flows, Organisational interdependencies, Authority-approval-supervision patterns, Inter-jurisdictional transaction patterns, Informal political decision chains

Source: Rondinelli (1985, p.143)

"development must be fitted to ecological constraints; priority attention (in agrarian economies) must be given to rural development; and planning for rural development must be decentralised, participatory, and deeply immersed in the particulars of local settings" (Friedmann and Douglas, 1978, p.163). In order to achieve such ends they advocate that national development strategies should be reoriented to include the following six policy elements: (i) the replacement of generalised wants by limited and specific human needs as the fundamental criterion of development; (ii) the treating of agriculture as a propulsive sector of the economy; (iii) the allocation of high priority to the attainment of self-sufficiency in domestic food production; (iv) the reduction of inequality in income and living conditions between social classes and between urban and rural areas; (v) the allocation of high priority to the production of wage goods for domestic production; and (vi) the adoption of planned industrial dualism involving the protection of small-scale production of the domestic market against competition from large-scale capital intensive enterprise. Each of these elements can be seen to have important repercussions for urban-rural interaction if they are incorporated into any planning framework.

Turning more directly to the implications of these arguments for the linkages between rural and urban areas, Stöhr (1981) has identified four keys areas where changes need to be introduced in the balance between the two if development from below is to be successful. At a political level he suggests that rural areas need to be given a higher degree of self-determination so that the flow of political power becomes less directly urban to rural in nature. Secondly, he suggests that national pricing policies should be introduced which offer terms of trade more suitable to agricultural and other rural products. Likewise productive activities in rural areas should be encouraged to exceed regional demand so that a pattern of export flows is generated. Finally, he suggests that the whole transport and communication network should be reorganised, both between urban and rural areas, and also at a village to village level. Bottom-up approaches such as these therefore seek to change the balance of a perceived inequitable flow of resources from rural to urban areas through "integrated regional resource utilization at different spatial scales" (Stöhr and Taylor, 1981, p.472).

Allocation and Access to Resources

The approaches of Lipton, Rondinelli, and Stöhr and Taylor are all concerned with the ways in which urban-rural relationships influence the nature of development, and in particular, with how these relationships determine the allocation of relatively scarce resources within society. In so doing they have highlighted two broad issues.

Firstly, while it is not always possible to distinguish clearly between urban and rural *people*, there are nevertheless a number of different types of linkage between towns and the countryside which can provide the basis for the measurement of urban-rural interaction. Using these it is then possible to produce a balance of flows for particular case studies, which can be used to indicate a net transfer of resources and an assessment of the degree of exploitation present. However, it is extremely difficult to interpret the results of such an approach to the study of rural-urban interaction, since it is *people* rather than *places* that exploit and are exploited. In Dixon's (1987, p.vi) words, "it is not urban and rural *areas*, or rural and urban *classes*, which interact". It is also not possible simply to speak of 'rich towns' and 'poor rural areas' - poor people and rich people live in both. What *is* interesting, however, is to see how different people and classes benefit from different types of urban-rural interaction, and thus how the flows between the two types of area are related to broader social and economic transformations. It is these relationships that the ensuing chapters seek to explore.

The second feature highlighted in the above discussion is the great diversity of factors influencing the direction of flow of resources between town and countryside. Although these are usually seen as being economic, as a result of differing terms of trade and pricing policies, it is important to emphasise that political, social and ideological factors also play important parts. This was well illustrated by the set of factors identified by Rondinelli as being of importance in determining whether cities were exploitative or

25

developmental, when he included such aspects as the aggression of local leaders and the willingness of people to accept social change. However, other factors, such as people's perceptions, particular development ideologies, and different class interests will also influence the direction of resource flows, and must be included in any analysis of urban-rural interaction.

Linkages, Flows and Interaction

Rondinelli's (1985) classification of linkages in spatial development, illustrated in Table 1.1, provides a broad framework for the consideration of urban-rural interaction. However, it conflates different kinds of *linkages* and likewise different kinds of *element*. He thus classifies population movement linkages as a separate category from physical and social interaction *linkages*, and he includes 'systems', 'networks', 'patterns', 'flows' and 'interdependencies' within his general *elements*. There is little clear logic in such a classification. Other authors have listed various types of flow and linkage between rural areas. O'Connor (1983) has thus suggested that in Africa the most important urban-rural linkages involve movements of people, the transmission of ideas, flows of goods and transfers of cash, while Gould (1985) has suggested that urban-rural interaction is the two-way flow of people, goods, money, technology, information and ideas. The types of flow or interaction studied depend partly on the theoretical framework within which research is based. However, there have been few attempts in the literature to distinguish between different types of linkage or interaction; in general terms, the words 'linkage', 'flow' and 'interaction' are used interchangeably.

In order to clarify these different terms, Table 1.2 presents a broad overview of urban-rural relations in which linkages, flows and interaction are seen as separate, but closely linked concepts. At the broad theoretical level it suggests that there are economic, social, political and ideological *linkages* between urban and rural places. These find their physical expression in measurable *flows* of, for example, people, money and budgetary allocation. These flows are associated

26

Table 1.2: Urban-rural linkages, flows and interaction

Linkages	Flows	Interaction
Economic	- labour - money - food - vehicles - commodities - energy - credit - raw materials	- labour/capital - marketing - shopping - transport
Social	- people - correspondence - telephone calls - medicine	- social groups - family - friends - class
Political	- power - authority - budgetary allocation - law	- political action - lobbying - justice provision - allegiance payments
Ideological	- ideas - books - radio - television	- religious activity - education - advertising

with *interactions* between people, places and objects, but do not in themselves actually embody those interactions. Thus labour flows are necessary for there to be an interaction between labour and capital, but they do not in themselves embody that labour/capital interaction. Table 1.2 only provides a few examples of urban-rural flows and interactions, but it does offer an overall framework within which to view the ideas, evidence and arguments put forward in this book.

Individual authors have concentrated on different aspects of these urban-rural relationships, focusing on a range of countries from the Caribbean, Africa and Asia. Morgan pays specific attention to *energy* flows in tropical Africa, and then considers the broader significance of these in terms of social and economic linkages. In so doing, he explores the ramifications of urban biomass fuel demand for rural producers of woodfuel. Sutton's chapter returns to some of the broader theoretical arguments concerning *urban bias*, and evaluates the rapid urbanisation and relative neglect of rural development that has taken place in the Maghreb in recent years. This is undertaken through an examination of agrarian reform, the technical development of water resources, and regional development policies in the region, all of which illustrate that increasingly centralised national space economies, often with growing regional disequilibria, have emerged in the countries of the Maghreb since their independence. Still within the African context, El-Bushra then provides a detailed case study of urban 'crisis' in the Sudan, and focuses attention on the question of *population* flows from peripheral rural regions to the Khartoum conurbation. An evaluation of this evidence suggests that heavy migration has resulted in a range of social and economic problems in both rural and urban areas. Simon also examines the influences of migration, but in his chapter he concentrates on the structural causes of *labour migration* flows within the very different political context of southern Africa. Drawing on the specific examples of Malawi, Mozambique and Zimbabwe, he illustrates how institutionalised labour migration, both within and between countries, has underlain the establishment of an integrated regional political economy based on the maintenance of poverty and inequality.

28

In Unwin's chapter, the regional focus shifts to the Arabian peninsula, where attention concentrates on the influences of recent urbanisation and industrialisation on rural-urban *food* flows. Particular emphasis is given to the neglect of agriculture in the development policies of the region's governments. Harriss, in a comparative study of Tamil Nadu in south India and Hausaland in west Africa also explores *food* flows in the context of grain commercialisation and the emergence of marketing systems. Following a theoretical assessment of the economies and social linkages associated with the development of market systems, she focuses attention specifically on the commercialisation of grain in her two case study areas, and the consequences of this for the nutrition of peasant households.

Corbridge provides a largely theoretical appraisal of 'counter-revolutionary' accounts of development, which see the 'Third World' as being beset by urban bias and excessive rural-urban migration. Drawing on examples from Asia and Africa, he suggests that the failure of neo-classical economics to address important questions concerning the nature of the 'market' in the Third World must cast a shadow over its supposed relevance to countries therein. He concludes that, although the 'counter-revolution' valuably emphasises the heterogeneity of developing countries, it tends to replace the concept of the 'Third World' with one of a monolithic market.

The final two case studies concern aspects of rural-urban interaction in the island states of the Caribbean. Potter emphasises the importance of the processes of colonial development and dependency from the fifteenth century in explaining the nature of subsequent social and economic change in small countries in the southern Caribbean. In a detailed empirical analysis he reveals clear differences between rural and urban areas in Barbados in terms of their provision of retail, commercial, health and other key facilities. These are shown to reflect past relationships between urban and rural areas, but they also continue to promote strong 'symbiotic'-parasitic *social and economic linkages* between such areas. Barker draws similar conclusions in his analysis of rural-urban relations in Jamaica. Here he emphasises the importance of the limited size of the island in influencing competition for scarce resources, the critical role of migration, and the significance of the environmental reciprocity between the rural interior and the coastal urban centres. In an historical assessment of the emergence of present rural-urban relationships he illustrates the way in which the rural economy became peripheral to the dramatic growth of the Kingston metropolitan core, and he concludes by arguing that environmental degradation can often amplify the ecological differences between rural peripheries and urban coastal areas in the region.

In the final chapter of the book, Potter draws together a number of arguments concerning rural-urban linkages, flows, interaction and development planning. In so doing, he stresses the importance of political paths to territorial development, citing in particular the experiences of Cuba. One over-riding conclusion that can be drawn from these examples is that it is essential to view urban-rural interaction as being the outcome of a series of underlying economic, social, political and ideological processes. These find particular expression in flows and linkages between rural and urban areas, and it is to be hoped that these essays will not only shed light on some of these processes, but will also encourage other scholars to turn their attention away from specifically *urban* or *rural* issues and to concentrate instead on urban-rural interaction in their exploration of change in developing countries.

Acknowledgements

I am particularly grateful to Rob Potter and David Simon for their comments on earlier drafts of this chapter.

Theoretical Perspectives

REFERENCES

Chambers, R. (1983) *Rural Development: Putting the Last First*, Longman, Harlow

Conroy, M.E. (1973) 'Rejection of growth center strategy in Latin American regional development planning', *Land Economics* 49, 371-80

Corbridge, S. (1982) 'Urban bias, rural bias, and industrialization: an appraisal of the works of Michael Lipton and Terry Byres', in: Harriss, J. (ed.) *Rural Development: Theories of Peasant Economy and Agrarian Change*, Hutchinson, London, 94-116

Dixon, D. (ed.) (1987) *Rural-Urban Interaction in the Third World*, Developing Areas Research Group, Institute of British Geographers, London

Friedman, J. and Douglas, M. (1978) 'Agropolitan development: towards a new strategy for regional planning in Asia', in: Lo, F.C. and Salih, K. (eds.) *Growth Pole Strategy and Regional Development Policy: Asian Experience and Alternative Approaches*, Pergamon, Oxford, 163-92

Gilbert, A. and Gugler, J. (1982) *Cities, Poverty and Development: Urbanisation in the Third World*, Oxford University Press, Oxford

Gould, W.T.S. (1982) 'Rural-urban interaction in the Third World', *Area* 14, 334

Gould, W.T.S. (1985) 'Rural-urban interaction in the Third World: building from R.U.I.N.', Department of Geography, University of Liverpool, mimeo.

Hansen, N.M. (1981) 'Development from above: the centre down development paradigm', in: Stöhr, W.B. and Taylor, D.R.F. (eds.) *Development from Above or Below? The Dialectics of Regional Planning in Developing Countries*, Wiley, Chichester, 15-38

Harriss, J. (ed.) (1982) *Rural Development: Theories of Peasant Economy and Agrarian Change*, Hutchinson, London

Harvey, D. (1985) *The Urbanization of Capital*, Blackwell, Oxford

Hoselitz, B.F. (1957) 'Generative and parasitic cities', *Economic Development and Cultural Change* 3, 278-94

Lipton, M. (1977) *Why Poor People Stay Poor: a Study of Urban Bias in World Development*, Temple Smith, London

Lipton, M. (1982) 'Why poor people stay poor', in: Harriss, J. (ed.) *Rural Development: Theories of Peasant Economy and Agrarian Change*, Hutchinson, London, 66-81

O'Connor, A. (1983) *The African City*, Hutchinson, London

Parr, J.B. (1973) 'Growth poles, regional development and central place theory', *Papers of the Regional Science Association* 31, 173-212

Potter, R.B. (1985) *Urbanisation and Planning in the Third World: Spatial Perceptions and Public Participation*, Croom Helm, London

Preston, D. (1975) 'Rural-urban and inter-settlement interaction: theory and analytical structure', *Area* 7, 171-74

Richards, P. (1985) *Indigenous Agricultural Revolution*, Hutchinson, London

Roberts, B.R. (1978) *Cities of Peasants: the Political Economy of Urbanization in the Third World*, Edward Arnold, London

Rondinelli, D.A. (1983) *Secondary Cities in Developing Countries: Policies for Diffusing Urbanization*, Sage, Beverly Hills

Rondinelli, D.A. (1985) *Applied Methods of Regional Analysis: the Spatial Dimensions of Development Policy*, Westview, Boulder

Santos, M. (1975) 'Underdevelopment, growth poles and social justice', *Civilizations* 25, 18-30

Stöhr, W.B. (1981) 'Development from below: the bottom-up and periphery-inward development paradigm', in: Stöhr, W.B. and Taylor, D.R.F. (eds.)

31

Theoretical Perspectives

Development from Above or Below? The Dialectics of Regional Planning in Developing Countries, Wiley, Chichester, 39-72

Stöhr, W.B. and Taylor, D.R.F. (eds.) *Development from Above or Below? The Dialectics of Regional Planning in Developing Countries*, Wiley, Chichester

Thomas, M.D. (1972) 'Growth pole theory: an examination of some of its basic concepts', in: Hansen, N. (ed.) *Growth Centers in Regional Economic Development*, Free Press, New York, 50-81

Wanmali, S.V. (1981) *Periodic Markets and Rural Development in India*, B.R. Publishing, Delhi

Zipf, G.K. (1949) *Human Behaviour and the Principle of Least Effort*, Hafner, New York

2

THE ROLE OF ENERGY IN URBAN-RURAL INTERCHANGE IN TROPICAL AFRICA

W.B. MORGAN

In a previous paper on urban woodfuel demand (Morgan, 1983) the author provided a comprehensive description of urban-rural relationships in energy use and supply in the Third World. The approach was, however, limited mainly to a discussion of the most important elements in the relationship, rather than to an examination of the wider social and economic context within which a fuller appreciation of policy implications might have been achieved. The paper was a contribution to a volume on woodfuel surveys, intended for persons confronted with the task of trying to deal with the growing fuelwood shortages, and it was thus intended more for those professionally engaged in forestry or agriculture than for policy makers or researchers. In consequence it did not examine the wider field of socio-economic development within urban-rural interchange or of rural transformation. Problems of energy use and supply cannot be satisfactorily divorced from those of food supply, income levels, the commercialisation of farming, rural-urban migration and the role of the city as the nexus of regional development. The use of woodfuel in Africa can only properly be appreciated in the context of poverty, of widespread income disparity and of the choice of alternative sources of energy. Nor can woodfuel use be divorced from the even broader context of worldwide economic depression and fluctuating world oil prices. Many of the more important aspects of woodfuel supply are not so much causes of economic problems as the symptoms of a

33

rural poverty which, as Wisner (1987) remarks, can only be alleviated, not cured, by the technological innovations and the forestry or agro-forestry prescriptions so widely suggested. They are better seen in the context of Chambers' (1983) integrated rural poverty.

So much of the rural or urban-rural research and literature concerned with the Third World has not only tended to isolate energy problems from other aspects of the economy and of society, creating an exaggerated impression at the same time of its importance, but has also done so with perceptions derived from an external, analytical and deliberately restricted view. For example, when the author was engaged in the United Nations University rural energy survey in Nigeria, he found that a number of respondents wanted the surveyors to put questions on other problems which they felt had a higher priority, such as water supply, and were disappointed when they failed to do so (UNU Rural Energy Systems Project briefly reported in Food and Agriculture Organisation, 1983, p.137-38). Chambers (1985) has referred to outsiders' biases, the biases of those who are neither rural nor poor, avoiding in many cases the worst seasons and the poorest people - for fear of giving offence - and restricting study to a particular specialisation. In Tanzania the Basic Needs Mission of 1980 was surprised by reports that there was no danger of deforestation since they supposed that this problem had been widely recognised at the national level. The villagers listed their five basic needs as food, water, dispensary, transport and above all essential commodities (International Labour Office, 1982, p.285; Chambers, 1985, p.8).

This chapter is an attempt to set the 1983 work in a broader context or at least to relate some of the features described to other aspects of social and economic change, so that the energy problem may be appreciated within the context of other priorities. In so doing, particular attention is paid to the role of energy in urban-rural interchange. The paper confines itself to tropical Africa, with a range of socio-economic conditions and of energy resources rather different from those of northern and southern Africa.

Urbanisation in tropical Africa

Although only a quarter of tropical Africa's population is urban, that urban population is nevertheless at the root of most economic and social change and, at the current annual growth rate of 6 per cent, will double in only 12 years. The rates of urban population growth by countries range from decline in Uganda at -0.1 per cent per annum in 1973-1984 to 11.3 per cent in Botswana (Figure 2.1) - in the rest of the world exceeded only by Oman, 17.6 per cent and Lesotho, 20.1 per cent. The data are subject to considerable probable margins of error and to variations in the definition of "urban" adopted in different African censuses. For sub-Saharan Africa, thus including southern Africa, the World Bank's Development Report for 1986 quotes an urban population growth rate of 6.1 per cent for 'low-income economies' in 1973-1984 (1984 GNP range per caput US$ 110-380) and 5.9 per cent for 'middle-income economies' (GNP range per caput US$ 450-1140). The urban proportions of population are shown in Figure 2.2, indicating a marked contrast between west, west-central and east Africa. Twenty-eight cities each have a population of over half a million persons - 20 years ago there were only three cities of this size. Of the urban population, 42 per cent in the 23 sub-Saharan African low-income economies in 1980 were in the largest city and 24 per cent in the 12 middle-income economies. Extremes of primacy were exhibited in Mozambique (83 per cent), Guinea (80 per cent), Senegal (65 per cent), Angola (64 per cent), and Benin (63 per cent). These relatively large nuclei in countries still retaining large rural populations and in many cases with economies still dominated by rural production, have enormously affected economic growth. As centres of political and economic power their pull on the surrounding countryside or backwash effect is clearly indicated in the population map showing great concentrations of rural settlement around the larger cities, in many cases with a geographical distribution resembling that of a gravity model, having a negative relationship between population density and distance from the city. The demographic interchange between town and country is clear, with all its implications for social and economic interchange. Whatever its legal and

Figure 2.1: Average annual growth rate of urban population, 1973–1983

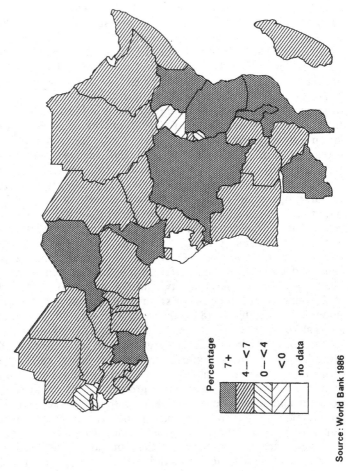

Percentage

7+

4 – <7

0 – <4

<0

no data

Source: World Bank 1986

36

Figure 2.2: Urban percentage of total population, 1984

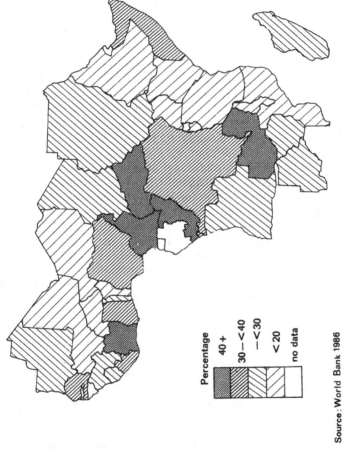

Percentage

40 +

30 — <40

— <30

< 20

no data

Source: World Bank 1986

37

managerial distinction, the city is not clearly separated from rural settlement and in tropical Africa many of its people belong both to the urban core and to the rural periphery, moving not infrequently between the two. For one aspect of energy distribution - the supply of woodfuel to the town - given that there is a high transport cost factor and a tendency to exploit the wood resources of nearby locations, the main woodfuel source lies in the more densely occupied farmland and some of the wood is transported through the regular movement of many people who are both urban and rural.

There is a high level of urban-rural interchange of resources and information, of which energy interchange is a part, although until recently this has been rather neglected in geographical research. Little work was done on biomass energy before the 1970s and even since then many authors have assumed that biomass fuels were almost entirely for rural subsistence and had very little commercial significance. Wood was treated as traditional or outside the modern sector of the economy, a dualistic approach criticised, amongst others, by Santos (1979), for it ignores the essential unity of all growth elements in African cities, both formal and informal, and the continuity between urban and rural areas of many elements of social life. It also ignores the commercialisation or commoditisation of rural production which puts a value not only on products sold to the town but also on those consumed in rural areas. It therefore tends to ignore the transformation of rural values and quality of life as a fundamental factor in commodity, including energy, interchange. Attention to the town as a focus of demand for biomass fuels or to urban-rural energy interchange in tropical Africa was drawn by Grut (1982), Digernes (1980), Foley and van Buren (1980), McGranahan *et al.* (1980), and Morgan (1983, 1985), although important indirect references have been made in other works such as Hosier (1985). No doubt there are other materials such as Omotara's undergraduate thesis, describing in detail the fuelwood retail network in a southern Nigerian town (Omotara, 1974).

Energy consumption, fuel costs and income

Urban areas are not necessarily the chief centres of energy consumption in tropical Africa, but they do consume the greatest variety of different kinds of energy and provide the greatest concentration of demand. Concentration makes possible a more effective market with the lowest prices for non-renewable fuels and electricity, and the cheapest systems of fuel and energy distribution. Towns are normally the chief consumers of electricity, butane gas, LPG (Liquid Petroleum Gas), kerosene or paraffin and, where available, coal or coke. In their central business districts and in their industrial and higher class residential districts high levels of energy consumption are associated with high incomes, the use of modern equipment, air-conditioning, lifts, street lighting, high levels of car ownership and a concentration of commercial transport. However, woodfuels are also consumed by all classes of residents and by many industries, partly because of certain cooking preferences with over 70 per cent of wood consumed in many African countries being for domestic cooking, partly because wood and charcoal can be stored and used as stand-by fuels when modern fuel and power supplies fail, and partly because of perceived relative costs. The last point has raised a number of problems in interpretation, however, as woodfuel costs per unit of heat produced often appear to be higher than those of other fuels. The general urban assumption is that woodfuel is used mainly by the poor, kerosene mainly by middle income groups, and electricity and gas mainly by those with higher incomes. For example, in towns with over 20,000 population in Sierra Leone in 1976-77 (Tables 2.1 and 2.2) more than 50 per cent of the incomes of the poorest people was spent on fuels and electricity, of which 59 per cent of energy expenditure was on woodfuel, 16 per cent on kerosene, 13 per cent on LPG and 12 per cent on electricity (Davidson, 1985). More was being spent on energy than on food. In the smaller towns and in the rural areas, wherever fuels had to be bought, a lower proportion of income was spent on fuels and electricity by the poorest people, but still over half of the expenditure on energy was on woodfuel. By contrast, the highest income group in the towns spent less than 13 per cent of its income on energy and in the larger towns bought

Table 2.1: Energy cost of total income in Sierra Leone, 1976-1977

Household income level (leones per month)	Percentage of fuel cost to income					
	Electricity	LPG	Kerosene	Fuelwood	Charcoal	Total
Urban population over 20,000						
0-100	6.27	6.81	8.08	15.12	15.59	51.87
101-200	5.37	3.87	3.46	9.92	8.62	31.26
201-400	4.13	2.16	2.75	3.08	4.24	16.46
> 400	4.66	1.88	1.64	2.12	2.29	12.59
Other urban population 2,000-20,000						
0-100	6.64	–	7.58	9.92	5.39	29.53
101-200	5.32	–	3.95	3.53	3.49	16.29
201-400	3.66	–	2.29	1.89	0.48	8.32
> 400	3.57	–	3.64	1.36	0.21	8.78
Rural	–	–	11.18	14.43	2.86	28.47

Source: Davidson (1985)

Table 2.2: Proportions of energy expenditure on different fuels by income groups in Sierra Leone, 1976-1977

Household income level (leones per month)		Percentage of energy expenditure on fuel				
		Electricity	LPG	Kerosene	Fuelwood	Charcoal
Urban population over 20,000	0-100	12	13	16	29	30
	101-200	17	12	11	32	28
	201-400	25	13	17	19	26
	> 400	37	15	13	17	18
Other urban population 2,000-20,000	0-100	22	-	26	34	18
	101-200	33	-	24	22	21
	201-400	44	-	27	23	6
	> 400	41	-	41	16	2
Rural		-	-	39	51	10

Source: Davidson (1985)

mainly electricity together with nearly equal proportions of the various other fuels. In the rural areas income groups were not differentiated, but outside the towns in most African countries incomes are extremely difficult to estimate - see for example Hosier's (1985, p.93-103) use of Crawford and Thorbecke's subjective income categories (Collier and Lal, 1980).

No table showing comparative costs by units of heat provided was supplied, but attention was drawn to the relatively high cost of kerosene, especially in the rural areas, and the high costs of fuelwood for the urban poor, reflecting not only transport costs but an increasingly important scarcity factor with rapid urban expansion and increasing distance to source of supply. Fluctuations in fuel prices of which there have been many, associated with the prospect of fuel prices rising much faster than people's incomes have very serious implications for standards of living, including levels of nutrition. Fluctuations may also cause people to be cautious about changing to a new, perhaps only temporarily cheaper fuel, especially if some capital expenditure is also involved. Digernes (1977) examined fuel expenditures by income groups in Bara, a small town in Sudan (Table 2.3). Again the lower income group spent the highest proportion of its income on fuel and the highest proportion of its energy expenditure on woodfuel, mainly charcoal brought from distant locations in a situation of local wood scarcity.

A number of analyses of fuel costs have been made in Africa of which three are cited in Table 2.4. Example A is from rural Kenya and shows the expected low costs per effective Mega-Joule for gathered fuelwood and for charcoal, higher costs for purchased fuelwood and higher still for paraffin or kerosene. Costs for gathered wood used on the farm have to be based on notional values of labour expended, which may depend on the estimated value of the alternative work available or the value set on leisure. Examples B and C are both urban and suggest that fuelwood is more expensive per unit of useful heat or delivered energy than other fuels or electricity. In both cases kerosene is cheapest, and in Nigeria it has been subsidised. In the 1980s, in many African countries kerosene prices have been dramatically reduced as a consequence of falling world oil prices. These examples serve merely as indicators of a highly variable pattern of fuel costs. A study in Malawi, for

Table 2.3: Monthly energy consumption and expenditure by income group, Bara Village, Sudan, 1977

	Low income			Medium income			High income		
	KCal	L	%FE	KCal	L	%FE	KCal	L	%FE
Kerosene	85.48	2.71	35	154.67	4.90	48	207.55	6.57	51
Wood	517.31	1.65	22	552.24	1.77	17	689.12	2.19	17
Charcoal	1608.75	3.30	43	1759.50	3.60	35	1998.38	4.09	32
Total	2211.54	7.66		2466.41	10.27		2895.04	12.85	
Estimated monthly average									
cash income		17.00			-			60.00	
Fuel expenditure as									
percentage of income		43.00			-			21.00	

Note: L – Sudanese pound; %FE – percentage fuel expenditure

Source: Digernes (1977)

Table 2.4: (A) Estimated cost per effective Mega-Joule cooking energy
(rural consumers in Kenya)

Fuel	Unit	Mark. price Ksh./unit	Calorif. value MJ/unit	Efficiency %	Cost/Effective MJ Ksh./MJ
Fwd gath.	kg	0.05	15.5	10	0.03
Fwd purch.	kg	0.20	15.5	10	0.13
Charcoal	kg	0.60	32.4	25	0.07
Kerosene	l	3.20	39.6	40	0.20

Source: 1981 Beijer Institute Survey, Kenya, in Hosier (1985, p.112)

Table 2.4: (B) Comparative prices of urban cooking fuels in Nigeria

Fuel	Purchase price Unit	k/unit	Ener. cont. MJ/unit	End use effic. %	Effective price k/MJ useful heat
Fuelwood	kg	17	15	8-13	8.7-14.2
Charcoal	kg	22	25	20-25	3.5-4.4
Kerosene	l	10	35	30-40	0.7-1.0
LPG	kg	34	48	45-55	1.3-1.6
Electricity	kWh	6	3.6	60-70	2.4-2.8

Source: UNDP/World Bank, 1983, Nigeria, in Leach (1987)

Table 2.4: (C) Costs of cooking fuels in Ouagadougou, Upper Volta (Burkina Faso), 1980

Fuel	Unit price CFA/kg	Calorific value MJ/kg	End use efficiency estimated %	Cost deliv. energy CFA/MJ
Fuelwood	15.0	14.65	6	17.0
Charcoal	50.0	33.14	15	10.0
Kerosene	137.5	42.90	50	6.4
Butane gas	256.0	45.02	55	10.3

Source: Direction Générale de la Recherche Scientifique et Technologique, Upper Volta, in Cecelski (1983)

example, has shown that in the early 1980s wood was much dearer in Blantyre than in Mzuzu, not only because wood in Blantyre came from much farther away, but also because Mzuzu households typically bought wood by the headload or bundle, and thus in bulk, whilst Blantyre households bought it in split pieces (Energy Studies Unit, Malawi, 1984, p.13 and p.15). If woodfuels are the fuel of the urban poor then in a number of instances the poor have apparently to live at a higher cost per unit of consumption than the rich. This phenomenon is not uncommon in other contexts in Africa. It can be seen in markets where the poor cannot afford to buy in bulk and where petty traders offer goods in extremely small units, such as single cigarettes and flour by the cigarette tin measure (Morgan and Pugh, 1969, p.575-77). In Mozambique in early 1986 charcoal cost 60-120 metical per kilogramme in 40 kilogramme sacks and 335-670 metical per kilogramme in 300 gramme tins (Marleyn, 1986a). Leach (1987) suggests that the poor use the most expensive fuels on the basis of useful energy and that useful energy falls in price as one moves from the inefficient woodfuels and equipment used predominantly by the poor to those used by the rich. He concludes that fuelwood costs compared with those of petroleum fuels seem to be the least important consideration. The rich cook much more efficiently with fuels that are not that much more expensive. For the African poor woodfuels have the advantage that they can be bought in small affordable quantities. This can be true of kerosene where the purchaser possesses a container capable of being sealed against evaporation, but is not true of butane or LPG. Woodfuels are also familiar fuels, for many of the urban poor are immigrants from rural areas where they are used to cooking on wood or charcoal fires and prefer the taste of food so cooked. Fuelwood can be burned anywhere in the open without extra cost. Charcoal can be burned in a very cheap stove often locally made, whilst kerosene, butane, LPG and electricity require relatively expensive cookers at a cost beyond the means of the poor. Comparative costs for stoves estimated by Floor (1977) in Burkina Faso and cited in DEVRES Inc. (1979) are shown in Table 2.5. For the poor the main use for kerosene and electricity is for lighting. However, so-called effective or delivered energy units are a useful measure only in ideal terms. In reality fire or stove

Table 2.5: Comparative costs for stoves in Burkina Faso

Fuel	Equipment	Cost of equipment US$	Price kw US$
Fuelwood	stones	0.00	7.2
Charcoal	stove	0.80-4.00	3.8
Kerosene	stove	10.00-14.00	4.4
Butane	stove & bottle	approx. 52.00	7.9
Electricity	stove & bottle	approx. 120.00	14.8

Source: DEVRES Inc. (1979)

Table 2.6: Charcoal prices in Angola

	Prices per kilogramme (kwanza)
to producers	125
transport	up to 40
by retailers	250-375
by consumers	800-1200

Source: Marleyn (1986b)

management is in itself a cost factor, and all forms of fuel use require experience for low-cost results. The introduction of new fuels or stoves may fail because the users fail to learn how to master the techniques required. Modern fuels are not just the energy of higher income groups, but generally also of better educated and more urbanised people. Africa's rural cooking often consists of slow simmering in large round-bottomed pots for a large family unit. Modern cooking in flat-bottomed metal pots and pans on stoves is generally more suited to the needs of the small nuclear family, but may become necessary because of urban living conditions, such as a one- or two-roomed apartment with limited ventilation and little or no storage space for wood. Falling kerosene prices in Africa suggest that in urban conditions there should be an increased use of kerosene stoves, possibly locally manufactured, and interest in kerosene substitution for wood is reviving (Foley, 1987). Inyang (1986) notes a 27-fold increase in domestic kerosene consumption in Nigeria from 1950 to 1982 and suggests that supply has been relatively dependable. He notes that three out of four households in Lagos occupy one-roomed apartments. However, experience with regard to the dependability of kerosene supply in Africa is highly variable.

Woodfuel burned in relatively confined spaces is frequently less clean to use and can involve health problems. Cecelski (1984) notes high emissions of suspended particulates and benzopyrene in biomass-using kitchens with urban standards, compared with general levels of urban pollution and the WHO recommended levels. Elsewhere in the Third World, the problem of high emission levels has been discussed by Aggarwal, Dave and Smith (1983).

The urban-rural energy interface

Overwhelmingly in tropical Africa commercialised energy is restricted to the urban population. Even in the Ivory Coast, Kenya and Zimbabwe where there are rural electric power distribution programmes, only a minority of people benefit. In West Africa, as Lambert (1984) notes, electric power consumption in rural areas is virtually non-existent. We might expect that the less urbanised countries would have a correspondingly greater dependence on woodfuel. The consumption of woodfuel by countries in tropical Africa as a percentage of total energy consumed (small quantities of energy other than woodfuel and commercial energy have been omitted) is shown in Figure 2.3. The highest percentages are in the poorest and least urbanised states, but high percentages also occur in relatively highly urbanised states like the Central African Republic and Congo and in states with relatively high GNP like the Ivory Coast. Analysis of the data for the urban percentage of population and the percentage of woodfuel consumed gives a correlation coefficient of -0.4146 (significant at over 0.4 for 31 degrees of freedom and a probability of 0.05, but not at p=0.01), suggesting the possibility of some small negative relationship between the two.

However, despite their use of modern commercial energy, the towns are also important users of biomass fuels from the rural areas and draw on the rural resource base for food, industrial raw materials, labour and even some cash and capital through investment and financial support for migrants to the towns who may take many months or more to find work. As towns grow the net flow of cash is almost certainly from urban to rural locations as ageing farm populations come to depend more and more on remuneration from the higher incomes of their urban-based children. The flows of fuel and electricity create an urban-rural energy interface, especially marked between kerosene and wood as each penetrates the other's base, but there are few cases of major penetration of rural areas by other fuels which can be regarded as a part of the general failure in Africa of the so-called spread effects (Morgan, 1983, p.54).

Woodfuel moves into towns mainly as the fuel of poor people living in overcrowded apartments in the inner city or in shanty towns or informal housing development on the urban periphery. Woodfuel is also supplied in smaller amounts to industry and institutions as charcoal or as split logs for furnaces and large ovens, and to the higher income groups as roundwood and charcoal. The concentrated market promoted by urban demand clearly puts pressure on nearby farmland and woodland resources, even acting as a species selector as different prices are offered for different qualities of wood determined by such factors as heat obtained,

Figure 2.3: Wood energy as a percentage of total energy (wood plus commercial energy), 1984

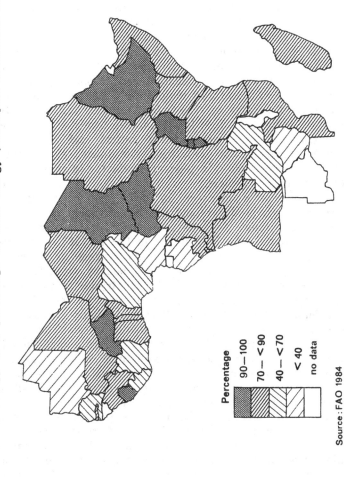

Percentage

90—100

70 — < 90

40 — < 70

< 40

no data

Source: FAO 1984

tendency to smoke and tendency to spark. This selectivity may result in the elimination of some species whilst the landscape may still seem well wooded. As wood prices rise so wood becomes more and more attractive as an income source, especially if food prices have been held down, either by government controls or by food imports, in some cases subsidised in order to reduce costs of living. Woodfuel sales can also be attractive to older farmers as a less demanding, even less risky way of making a living than crop production or as a supplement to crop sales, often not taxed. Roundwood has a special importance as it can be produced by pollarding and its production can be maintained from year to year. In rainforest regions old cocoa and coffee trees make good sources of fuelwood, but for each farm the resource is limited. Split logs from tree trunks are produced by gangs of professional wood cutters and are usually supplied directly to industrial consumers by contract, whereas roundwood is a peasant product sold in small lots through rings of dealers.

A considerable mark-up in price can occur between purchase from producers and sale to consumers as the wood may pass through four or five stages in the distribution system. Marleyn (1986b) quoted in Leach (1987, p.36) cites the mark-up of charcoal prices in Angola as shown in Table 2.6. Roundwood is also bought directly from producers by urban visitors to the countryside or transported as a make-weight load, bought and sold by lorry drivers to supplement their income. Woodfuel transport costs are complicated by the frequent combination of wood loads with other goods, whilst the production costs are complicated by the combination of fuelwood with other products such as timber and crops, and by the sale of wood as a by-product of land clearance or timber cutting or as the result of pollarding or thinning. The two sets of costs are interrelated. One other factor is the question of a return load from the town to the rural area. Mortimore and Wilson (1965), for example, drew attention to the movement of night soil out of Kano in northern Nigeria as a return load on the donkeys which had brought in crops and fuelwood from the surrounding farmlands.

Charcoal with its much greater heat value per unit weight can considerably extend the urban woodfuel supply area. In

Senegal, for example, the Dakar supply area extends for over 300 kilometres, very much more than would have been possible with wood transported as a normal commercial load. Charcoal can reduce pressure on farmland close to the town, although as distance to source increases and prices rise, small scale charcoal production in low cost pit kilns on farms near the town can be remunerative despite their low efficiency. Such kilns also have advantages in areas of dense woodland in that charcoal manufacture can always move to the trees and need not depend on plantations. Charcoal is particularly important as an urban domestic fuel in East Africa but rather less so in the west (Figure 2.4). One might have expected some relationship in tropical Africa between charcoal consumption and percentage of urban population, but the data fail to support it. One might also have expected that charcoal use would increase as woodfuel dependence was reduced. If we take the nine countries with the highest percentages of woodfuel consumption in charcoal and compare with wood energy as a percentage of total energy consumption (Table 2.7), we can see that some of them, such as Chad, Sudan and Sierra Leone, have very high levels of wood energy dependence, whilst others, such as Liberia and Zambia have much lower levels. In several countries, such as Kenya, charcoal is also used in rural areas, where its importance can increase wherever real incomes are rising and the opportunity cost of rural labour is increasing or where labour shortage occurs and people are less willing to spend time gathering fuelwood (Hosier, 1985, p.113-114). Small kilns are widespread in East Africa, but in Senegal burners' cooperatives use large earth kilns (Foley and Van Buren, 1980). In Zambia there is a charcoal burners' union and their wood cutting has to be licensed and restricted to designated areas (Kay, 1967, p.94). Pressure on biomass resources may also encourage the use of alternative materials such as crop and timber wastes. In Nigeria sorghum stalks have long been used for fuel, even in the towns, whilst oil seed wastes are commonly used in West Africa as fire lighters (Morgan, 1983). Briquetting wastes for use as domestic and industrial fuel is a growing industry in several countries (Bennett, 1987; Leitmann, 1987; Louvel, 1987).

The kerosène distribution system provides a network along which other modern fuels and energy innovations may flow.

Figure 2.4: Charcoal as a percentage of woodfuel consumption, 1984

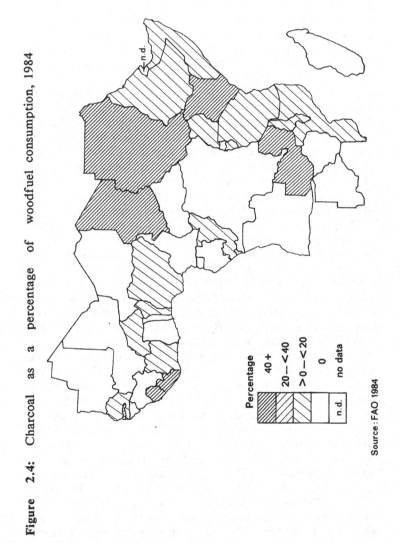

Percentage

40 +

20 — <40

>0 — <20

0

n.d.   no data

Source : FAO 1984

Table 2.7: The leading charcoal consuming countries of Tropical Africa in 1984

	Charcoal as a percentage of woodfuel consumption	Wood energy as percentage of total energy consumption
1 Chad	73	91
2 Sudan	69	78
3 Sierra Leone	67	91
4 Liberia	48	62
5 Zambia	46	59
6 Kenya	40	84
7 Gambia	30	75
8 Senegal	17	52
9 Ivory Coast	13	57

Kerosene cooking is limited in its rural distribution to those families whose labour has high opportunity costs or who otherwise lack wood. However, it should be seriously regarded as a possible alternative to wood in those areas of the African Sahel where wood shortages have reached alarming levels and where environmental damage has occurred, although many locations are admittedly long distances from port and transport costs are high. Throughout Africa the dispersed nature of the rural market makes the distribution of goods from urban sources extremely expensive. Wood is the ideal dispersed fuel to satisfy rural needs. However, the high population densities of southeastern Nigeria, for example, are much less costly to supply than the low densities of peripheral Mali, and the grouping of population into large villages may also reduce distribution costs.

Oil products are important in rural areas not only as a source of domestic energy but also as transport fuel and a potential source of power on the farms. Likewise, their importance as a source of industrial power has an effect on the economy with very serious implications for agriculture and rural prosperity. We can distinguish two groups of African countries: the oil exporters with a commercial energy surplus and the oil importers with a commercial energy deficit (Figure 2.5). After 1973, the deficit countries were affected by the rise in commercial energy costs, reducing the amount of foreign exchange available for other imports, including the materials and equipment needed for investment programmes. Some reduced the rate of development and reduced oil expenditure, whilst others with more limited development found themselves unable to reduce oil imports any further (Morgan and Moss, 1980, p.8-11). In the rural areas the cost of moving farm produce and woodfuel to market rose. Woodfuel sources near the towns became more attractive and charcoal's transport cost advantage was even more appreciated. As kerosene prices rose so did wood prices, although in some cases energy prices were fixed by government. For example, the Senegal government held charcoal prices constant from 1975 to 1980 (Foley and Van Buren, 1980). The plight of the urban poor was made worse as fuel costs absorbed more of their income, while more distant rural locations were threatened with greater isolation.

Figure 2.5: Net exports and imports of commercial energy in Tropical Africa, 1983

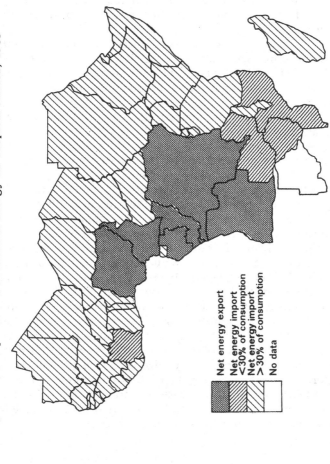

Net energy export

Net energy import
<30% of consumption

Net energy import
>30% of consumption

No data

Source: U.N. Statistical Yearbook 1985

In many African countries food imports continued to increase during the 1970s and combined with oil in absorbing large shares of foreign exchange income. A comparison of the 1965 and 1983 shares of merchandise imports (Table 2.8) illustrates the problem. A few countries like Benin managed to achieve small reductions in the food and fuel share of the value of imports, and others like Nigeria with its own oil production managed to reduce the share of fuel imports but experienced an increase in the share taken by food. In some countries food imports remedy problems of food deficiency, but in others, food imports are a cause of poor agricultural productivity rather than an effect.

In the oil exporting countries, 1973 to 1981 saw a sharp increase in oil's share of GDP. In Nigeria expenditure on international tradable and non-tradable goods rose, but resources were attracted out of the internal non-oil tradable sector - mainly agriculture (Collier, 1987). Export crop production declined, but manufacturing production rose, due partly to the favourable non-market resource allocation - from 1973 to 1981 at 9.4 per cent per annum in real terms (Jazayeri, 1986). In 1975-1979, agriculture received only 7.1 per cent of public investment (9.9 per cent if irrigation is included) and export crop production suffered from a mass exodus of labour (Jazayeri, 1986). Real agricultural output was estimated to have fallen by 2 per cent per annum. Apart from countries like Malawi and Cameroon, many other African countries protected manufacture, taxed agricultural exports and tried - not always successfully - to control food prices (see Agarwala, 1983). After 1981, as world oil prices fell, there was no revival in Nigeria of the non-oil export sector. By 1985, real wages in the public sector fell by 40 per cent and many urban workers returned to the rural areas. Consequently subsistence food production and the use of fuelwood in the rural areas inevitably increased.

Land and labour

Whilst woodfuel had been treated simply as a subsistence product or as the by-product of forest clearance, it was regarded as outside normal economic considerations. The

Table 2.8: Percentage share of the value of merchandise imports in food and fuels in selected countries of Tropical Africa in 1965 and 1983

	1965 Percentage of imports		1983 Percentage of imports	
	Food	Fuel	Food	Fuel
Benin	18	6	16	5
Burkina Faso	23	4	23	17
Cameroon	11	5	9	4
Congo	15	6	17	15
Ethiopia	6	6	9	25
Ivory Coast	18	6	20	19
Liberia	17	8	25	17
Madagascar	19	5	16	24
Nigeria	9	6	21	3
Sierra Leone	17	9	27	35
Zambia	9	10	9	19

growing realisation of the fact that there are markets for woodfuel in tropical Africa has prompted a sharper focus on production and on the organisation of the woodfuel industry which is seen in many cases to incur both land and labour costs. Hosier (1985, p.103-104) comments that fuelwood in Kenya appears "to be one of the last goods to be commoditised, that is to say, universally bought and sold on a market". In the rural areas it has a use-value rather than an exchange value, but he concludes that some fuelwood is purchased in rural Kenya, some households pay for the right to cut trees and collect deadwood on private land, and charcoal is commoditised with increasing demand due to the processes of rural transformation and urbanisation. Its production is more and more "liable to be undertaken by large firms seeking a profit" (Hosier, 1985, p.104). Wisner (1987) goes further and writes of the acceleration of the private ownership of land in Kenya since independence, although private tenure with excessive subdivision and fragmentation has long been a feature of the Kikuyu lands (Bohannan, 1963). The process of adjudication of usufructuary rights has been completed on all land of high or medium productive potential and through this process a new landed class has arisen. Those with large holdings not wholly in crop production, can derive profit from timber and woodfuel, especially charcoal. On very small farms the rural poor 'over-exploit' their plots both for crops and fuel and the pattern of land ownership is beginning to create localised areas of vegetation deterioration as the poor are forced into adopting short term solutions to their financial problems. Wisner (1987) refers to women working more than twenty hours a week collecting fuel, but notes that with major difficulties in providing food, clothing and shelter, domestic energy is not the primary problem although in the long-run deforestation could have a profound effect on the viability of the entire rural economy.

Kenya is an extreme example of the development in tropical Africa of private ownership in land. In other states more emphasis is put on communal or socialist forms of tenure. Nevertheless trees are acquiring commercial value almost everywhere, even trees not bearing marketable fruit or valued for timber. On common land where access to fuel is freely available, a rural market in woodfuel cannot exist, but in some cases, as in southern Nigeria, "crops, trees and houses are not considered as inseparable from the soil on which they stand and may be owned by individuals" (Morgan and Pugh, 1969, p.72). Trees may be marked by owners after girdling or killing with chemicals so that ownership during the drying out period is clear.

A great deal of woodfuel is gathered by women and children in their 'spare time', but the task is becoming more demanding. Müller and Nivelle (1970) referred to women and children in Togo walking 20 to 25 kilometres in a day to obtain just over 1/8 stere of wood (approximately 45 kilogrammes of dry wood). In the Sahel, improved woodfuel use and piped water could make more time available for food processing and preparation (Sokona *et al.*, 1986). In some countries wood collection has been the principal activity of children (Cecelski, 1984, p.19-21), but has been affected by the spread of primary education which has reduced the amount of child labour time available. Digernes (1977) has recorded the role of children in Bara, Sudan, as cutters, transporters and sellers of wood. The importance of their role as family labour has been cited by Cecelski (1984) as a contributory cause of high birth rates. As wood has become marketable, wood earnings have become part of the farm income, together with the earnings from other saleable gathered products. Part of the woodfuel supply depends on the employment of paid labour, chiefly for cutting split logs and for large-scale charcoal production. As woodfuel becomes commercialised so the prospects improve for the development of agroforestry for timber, woodfuel and crops. However, many locations in Africa suffer from land tenure constraints, shortage of willing labour, alternative demands on labour time or from vested interests in the existing woodfuel industry. In Nigeria's oil boom the decline of agricultural production and the reduction of the rural labour force helped locally to raise fuelwood prices.

As woodfuel becomes commercialised, the income from it tends to be small at first, so that woodfuel collection and selling are often regarded as lowly occupations. In southwestern Nigeria the sellers have tended to be over 35 years of age as younger women tended to find the trade degrading (Morgan, 1978). In Ghana, whilst women traditionally collect fuelwood, men work in charcoal production, but the decline of Ghana's market economy in the 1970s reduced wage opportunities, forced families back into subsistence production and encouraged more men into fuelwood gathering and distribution (Cecelski, 1984, p.23). The rural labour demand is of course very seasonal. Woodfuel is most easily gathered and distributed in the dry season when it dries easily, travel is easier, labour is freer of other tasks and when most fallow and woodland clearance takes place. It gives farmers an income mainly in the mid or late dry season after harvest when otherwise they may have very little to sell. In the urban areas wood prices rise during the rains as the supply is reduced and many customers have very little storage space. In Senegal, for example, the price of charcoal has been known to double in the rainy season (Foley and Van Buren, 1980). More distant and drier sources of woodfuel become competitive, with the distance factor in woodfuel supply possibly fluctuating with the seasons, especially if local labour is in short supply because of the demands of agriculture.

Innovation

Woodfuel commercialisation has encouraged innovation and its diffusion. Some of this innovation is of urban origin, as with electrification, electrical equipment, kerosene and butane stoves, and such innovation has spread outwards from the city. Other innovation, whatever its origins, is more rurally based as an intermediate or alternative technology more adapted to the needs of dispersed people with low incomes, although such technology is also targetted on urban needs, more especially on the needs of the urban poor. Some attempt has been made to model the space relationships of the different innovations (Morgan, 1983) but the scheme had little practical relevance partly due to the unsuitability or relatively high cost of most of the technology available. The attempt to develop intermediate energy innovations is now a world-wide industry responding to rising fuel costs, changing social conditions and values, changing income levels, the urbanisation process and the perception of declining resources - especially woodfuel resources - accompanied by environmental degradation. Elsewhere the author has argued for a mixed energy development strategy (Morgan and Moss, 1980, p.201-203). Such a policy seems essential in order to use a wide range of material and labour resources, rather than risk the disequilibrium likely to result from concentrating on a single strategy. Bottled gas and LPG, like kerosene, can substitute for wood - Nigeria, for example, as a major oil producer, has fixed oil product prices for the home market at levels low enough to encourage substitution (Corvinus, 1983). Another possible substitution in the oil producing countries is natural gas, currently flared off in quantities exceeding total energy consumption. If a gas pipeline network cannot be built, then the gas may be used to generate electricity and fed into a grid, or industrial development may be concentrated at least in part near the gas fields. Not only is the gas potential only partly developed, but so too is the hydro-electric power potential. There are several hydro-electric power schemes in tropical Africa, but so far they use only a very small proportion of the power which could be made available. The hope that falling world oil prices would result in lower kerosene prices and kerosene subsition for wood, as Foley (1987) has argued, has had only limited realisation, partly because of stove and distribution costs, partly because in many rural areas kerosene cooking is disliked or the stoves have proved unsuitable, and partly because of lags in retail price adjustments. However, in the towns the demand for kerosene has been rising, with, for example, the demand in Kenya more than doubling between 1970 and 1979 (Schipper, Meyers and Sathaye, 1984). Alternative energy substitutes have had only limited application in tropical Africa and so far have achieved very little success. Rotty (1980, p.889) has observed that alternative energy systems will be unable to supply most of the developing world's needs although the developing countries appear to provide the greatest opportunities for their application. Zimbabwe is one of the very few countries with a 'gasohol' (ethanol substitution for petrol) plant based on sugar cane production - the Triangle plant, which is the largest outside Brazil, constructed with the intention of achieving a 15 per cent level of petrol substitution. Coal and coke use is restricted to the few countries with coalfields, notably Mozambique, Zimbabwe,

Zambia, Zaire and Nigeria, and to ports, where in some cases it is used in power stations. Its use for domestic cooking is much more problematic, as Foley and Van Buren (1980) argue.

For most of tropical Africa the main concerns of energy strategy outside the industrial sector are either increased wood production or decreased wood consumption, usually through greater efficiency in wood use. Most policies usually aim at a combination of both. The urgency of developing an effective wood energy strategy is greatest in the drier countries, especially in the Sahel (discussed in CILSS, 1978). In the moister countries the woodfuel problems appear to be more long term, although in some, agricultural expansion involves high rates of forest destruction. However, it is possible that in many cases the rate of such expansion is decreasing as urban growth results in the development of a centripetal tendency in population movement. The signs of agricultural decline in Africa and increasing dependence on food imports support this thesis, although any major move towards self-sufficiency to counter economic decline, loss of exports and mounting international debts will reverse this trend. Both tendencies exist. In most tropical African countries the combination of growing woodfuel demand with rapid urban expansion produces the most severe environmental damage resulting from the rural energy industry and also sets a growing economic problem for the urban poor. Agroforestry schemes in the surrounding farmlands seem more likely to offer a solution than woodfuel plantations, for they will make use of an existing production system and labour force, although woodfuel plantations have been established near cities in several African countries since the last decade of the last century when the conservation of woodfuel and timber resources was seen as a major duty of the newly established forestry services. It is still an element of current energy policies as, for example, in Tanzania and in Senegal. In an agroforestry scheme wood products can be distributed through the existing petty trader system providing work for many poor people, whereas plantations may well bring a change in the scale of production and also in the size of the distribution units. Plantation programmes are similarly likely to be very expensive compared with either agroforestry or indigenous forest management. In

Malawi it was noted that only part of the likely cost could be recovered by wood sales (Energy Studies Unit, Malawi, 1984, p.32).

Estimation of woodfuel consumption for tropical African cities and the productive area required is extremely difficult. If one assumes a consumption of 1.0-1.7 cubic metre caput per annum (Delucia, 1983), an average weight of 0.725 tons per cubic metre and a productive capacity in savannas and rainforest under self regenerative conditions ranging from 0.073 tons to 3.77 tons per hectare per annum (Nigeria - see Morgan, 1978), then a city of 300,000 population satisfying 60 per cent of domestic demand with fuelwood would require anything from 30,000 to 3,000,000 hectares, combining the best and worst conditions for production with the heaviest and lightest demands. There could be serious limitations of land availability within the economic range of supply. There could also be a problem of training a sufficient labour force in the appropriate management skills. Even with the agroforestry alternative higher wood prices will be needed in many areas to provide sufficient encouragement to plant trees. Many village and school tree planting programmes have failed because of insufficient inducement or the lack of appreciation of the existence of a woodfuel supply problem.

At the same time the introduction of wood stoves or of more efficient stoves could reduce the rate of wood consumption. Wood stoves are usually much cheaper than kerosene cookers and many designs can be built very cheaply by the users themselves, using local materials. Cecelski (1984, p.78-86) has provided an excellent critique of stove programmes, their frequent failure under field conditions, the little regard of stove users for techniques of fuel saving and the preference of many cooks for smokelessness and time saving, which wood stoves often fail to achieve. Hosier (1985, p.135) also lists problems facing wood stove adoption including the fact that many of the 'improved' stoves are even less efficient than the traditional open fire. Other problems with wood stove introduction include loss or reduction of space heating, an inflexible cooking surface often lacking sufficient pot holes and loss of the smoke wanted to dry thatch and kill insects. A surface high enough for small pots may be too high for large ones, and the stove may require wood lengths different from those normally available. Some of these features conflict with one another. With any innovation the gains and the losses have to be weighed against one another. A critical but optimistic assessment of an improved wood stove programme has been produced by Ki-Zerbo (1980). Cecelski (1984) calls for greater user participation in stove design. Throughout the Third World work on stove design is gathering pace. Many of the achievements have been recorded in *Boiling Point*, the newsletter of the Intermediate Technology Development Group stove project. Many of the criticisms arise in the rural areas where fuelwood shortage may be seen as a less important problem than low incomes and water supply. In the towns where fuel can cost a third to a half of the incomes of the poor, reduced wood use clearly has a more immediate appeal, providing the new stoves can be assembled at very low cost and training schemes can be introduced to enable cooks to acquire the new techniques quickly or to adapt themselves where necessary to new lifestyles. The needs of changing urban economy and society and of growing pressure on the rural environment must surely make the effort worthwhile.

REFERENCES

Agarwala, R. (1983) 'Price distortions and growth in developing countries', *World Bank Staff Working Papers* 575, Management and Development Series 2, World Bank, Washington D.C.

Aggarwal, A.L., Dave, R.M. and Smith, K.R. (1983) 'Air pollution and rural biomass fuels in developing countries', *Asset* 5, 18-24

Bennett, K. (1987) 'Groundnut shell briquetting in the Gambia, *Boiling Point*, 12, 14-15

Bohannan, P. (1963) '"Land", "tenure" and land tenure', in: Biebuyck, E. (ed.) *African Agrarian Systems*, International African Institute and Oxford University Press, London, 110-115

Cecelski, E. (1983) 'Energy needs, tasks and resources in the Sahel: relevance to wood stoves programmes', *GeoJournal* 7, 15-23

Cecelski, E. (1984) *The Rural Energy Crisis, Women's Work and Family Welfare: Perspectives and Approaches to Action*, Rural Employment Policy Research Programme, World Employment Programme Research Working Papers, International Lab. Office, Geneva

Chambers, R. (1983) *Rural Development: Putting the Last First*, Longman, London

Chambers, R. (1985) *The Crisis of Africa's Rural Poor: Perceptions and Priorities*, Discussion Paper 201, Institute of Development Studies, Sussex

CILSS, (Comité Permanent Inter-Etats de Lutte Contre la Sécheresse dans le Sahel) (1978) *Energy in the Development Strategy of the Sahel*, directed by E.T. Ferguson, Club du Sahel, Paris, study carried out by Lahmeyer International GMBH, Frankfurt am Main, ORGATEC, Dakar and SEMA, Montrouge

Collier, P. (1987) 'Macroeconomic effects of oil on poverty in Nigeria', *IDS Bulletin*, Institute of Development Studies, Sussex

Collier, P. and D. Lal (1980) 'Poverty and growth in Kenya', *World Bank Staff Working Papers* 28, World Bank, Washington D.C.

Corvinus, F. (1983) 'Strategies for petroleum substitutes on the Malaysian energy market', *GeoJournal* 7, 41-52

Davidson, O.R. (1985) *Energy Use Patterns: Sierra Leone*, International Development Research Centre, Ottawa

Delucia, R. (1983) 'Defining the scope of woodfuel surveys', in: *Wood Fuel Surveys*: Forestry for Local Community Development Programme, Food and Agricultural Organisation of the U.N., Rome, 5-28

DEVRES Inc. (1979) *Socio-Economic and Environmental Context of Fuelwood Use in Rural Communities: Issues and Guidelines for Community Fuelwood Programmes*, US Agency for International Development, Washington D.C.

Digernes, T.H. (1977) *Wood for Fuel: Energy Crisis Implying Desertification. The Case of Bara, The Sudan*, Thesis, University of Bergen, Norway

Digernes, T.H. (1980) 'The energy situation of a town in the semi-arid Sudan', in: Morgan W.B., Moss, R.P. and Ojo, G.J.A. (eds.) *Rural Energy Systems in the Humid Tropics*, Proceedings First Workshop, UNU Rural Energy Systems Project, Ife, Nigeria, UN University, Tokyo, 21-25

Direction Générale de la Recherche Scientifique et Technological (1983) *Energie Nouvelles et Developpement* 12, 2-80

Energy Studies Unit, Malawi (1984) *Malawi Urban Energy Survey*, Ministry of Forestry and Natural Resources, Lilongwe, Malawi

Floor, W.M. (1977) *The Energy Sector of the Sahelian Countries*, Policy Planning Section, Minsitry of Foreign Affairs, Netherlands

Foley, G. (1987) 'The kerosene option', *Boiling Point*, 12, 20-22

Foley, G. and A. Van Buren (1980) *Coal Substitution and Other Approaches to Easing the Pressure on Wood Fuel Resources: Case Studies in Senegal and Tanzania*, Conference on New and Renewable Sources of Energy, 2nd Meeting of Technical Panel on Fuelwood and Charcoal, International Institute for Environment and Development, London and F.A.O., Rome

Food and Agricultural Organisation (1983) *Wood Fuel Surveys: Forestry for Local Development Community Programme*, F.A.O., Rome

Energy in Tropical Africa

Food and Agricultural Organisation (1984) *F.A.O. Yearbook of Forest Products 1984*, F.A.O., Rome

Grut, M. (1982) *The Market for Firewood, Poles and Sawnwood in the Major Towns and Cities in the Savanna Region*, Technical Report 6, F.A.O., Rome

Hosier, R. (1985) *Energy Use in Rural Kenya: Household Demand and Rural Transformation*, Energy, Environment and Development in Africa 7, Beijer Institute, Stockholm and Scandinavian Institute of African Studies, Uppsala

International Labour Office (1982) *Basic Needs in Danger: a Basic Needs Oriented Development Strategy for Tanzania*, Report to the government of Tanzania by a JASPA (Jobs and Skills Programme for Africa) Basic Needs Mission, I.L.O., Addis Ababa

Inyang, I.B. (1986) 'The energy "mix" for urban poor', *Work in Progress*, UN University, 10, 4

Jazayeri, A. (1986) 'Prices and output in two oil-based economies the Dutch disease in Iran and Nigeria', *IDS Bulletin*, 17, 14-21, Institute of Development Studies, Sussex

Kay, G. (1967) *A Social Geography of Zambia*, University of London Press, London

Ki-Zerbo, J. (1980) *Improved Wood Stoves: Users' Needs and Expectations in Upper Volta*, VITA (Volunteers in Technical Assistance), Mt. Rainier, Maryland

Lambert, Y. (with the collaboration of B. Meunier) (1984) 'Energy needs in West African countries: possible contribution of renewable energy', in: Lambert Y., Fall, D. and Isaza, J.F., *Renewable Energy Resources*, International Development Research Centre Energy Research Group, Ottawa, 5-31

Leach, G. (1987) 'Energy and the urban poor', *IDS Bulletin* 18, 31-38, Institute of Development Studies, Sussex

Leitman, J. (1987) 'Residue utilisation: a recent example from Africa', *Boiling Point* 12, 11-14

Louvel, R. (1987) 'Evaluation of briquette acceptability in Niger', *Boiling Point* 12, 15-18

McGranahan, G., S. Chubb and R. Nathans (1980) 'Patterns of urban household energy use in developing countries: the case of Nairobi', in: *Energy and Environment in East Africa*, International Workshop Sponsored by Beijer Institute, Keyna Academy of Science and UNEP, UNEP, Nairobi, 178-231

Marleyn, O. (1986a) *Field Report - Mozambique*, SADCC Fuelwood Project, ETC, Leusden, Netherlands

Marleyn, O. (1986b) *Field Report - Angola*, SADCC Fuelwood Project, ETC, Leusden, Netherlands

Morgan, W.B. (1978) 'Development and the woodfuel situation in Nigeria', *Geojournal* 2, 437-442

Morgan, W.B. (1983) 'Urban demand: studying the commercial organization of wood fuel supplies', in: *Wood Fuel Surveys: Forestry for Local Community Development Programme*, F.A.O., Rome

Morgan, W.B. (with the assistance of P.A. Moss) (1985) 'Biomass energy and urbanisation: commercial factors in the production and use of biomass fuels in tropical Africa', *Biomass* 6, 285-299

Morgan, W.B. and R.P. Moss (1980) *Fuel and Rural Energy: Production and Supply in the Humid Tropics*, Tycooly Press for the UN University, Dublin

Morgan, W.B. and J.C. Pugh (1969) *West Africa*, Methuen, London

Mortimore, M.J. and J. Wilson (1965) 'Land and people in the Kano close settled zone', *Occasional Paper* 1, Department of Geography, Ahmadu Bello University, Zaria, Nigeria

Müller, M. and Nivelle, J.L. (1970) 'Dévelopment des Resources Forestières: Togo, Economie Forestière et Perspectives, UNDP-FAO, Rapport Technique 2, Rome

Omotara, B.A. (1974) *The Supply and Distribution of Firewood in Ile-Ife*, B.Sc. Essay, Department of Geography, University of Ife, Ile-Ife, Nigeria.

Rotty, R. (1980) 'Growth in global energy demand and contribution of alternative energy supply systems', in: Bach, W. *et al.* (ed.) *Renewable Energy Prospects*,

UNU *et al.*, Conference Proceedings on Non-Fossil and Non-Nuclear Fuel Energy Strategies, UNU, Tokyo and Pergamon, Oxford, 881-890

Santos, M. (1979) *The Shared Space: the Two Circuits of the Urban Economy in Underdeveloped Countries*, Methuen, London

Schipper, L., Meyers, S. and Sathaye, J. (1984) 'Energy demands in the developing countries: toward a better understanding', in: *Energy Demand Patterns*, Energy Research Priorities Seminar, 1983, Energy Research Group, Ottawa

Sokona, Y. et al (1986) 'Women: food and energy providers of the Sahel', *Work in Progress* 10, 9, UNU, Tokyo

UNDP - World Bank (1983) *Nigeria: Issues and Options in the Energy Sector*, World Bank, Washington D.C.

Wisner, B. (1987) 'Rural energy and poverty in Kenya and Lesotho: all roads lead to ruin', *IDS Bulletin*, 18, 23-29, Institute of Development Studies, Sussex

World Bank (1986) *World Development Report 1986*, Oxford University Press, London and New York

3

THE ROLE OF URBAN BIAS IN PERPETUATING RURAL-URBAN AND REGIONAL DISPARITIES IN THE MAGHREB

KEITH SUTTON

The 'Urban Bias' nature of Post-Independence Socio-Economic Development

It can be argued that all countries of the Maghreb - Morocco, Algeria, Tunisia, and Libya - have suffered from 'urban bias' in their recent socio-economic development. Despite variations in their political systems, ranging from Tunisia's economic liberalism to Algeria's centralised command economy approach and encompassing Libya's idiosyncratic version of Islamic Socialism, all four North African countries reflect this 'urban bias' not only in their rapid urbanisation, but also in their related neglect of rural and peripheral development. Even where substantial programmes of rural or regional development have been implemented, it can be demonstrated that these are only palliatives relative to the general urban-industrial orientation of the development strategy or that, over time, well-intentioned efforts at rural or regional development have been deflected and diluted by the inherent centralist tendencies implicit in 'urban bias'.

One aspect of the Maghreb's colonial 'legacy' was the initiation of rural-urban flows of agricultural rents, profits and produce to the burgeoning towns. These towns, often also ports and frequently pre-colonial in origin, were enormously enhanced in status and function as the prime links in the nineteenth-century imperialist system which enmeshed the formerly rather isolationist Maghreb region into the European economic system. They became the focus for the main areas of European settlement, especially around

Oran, in the Mitidja Plain near Algiers, and in the Medjerda Valley close to Tunis. These and other coastal towns like Bône and Philippeville developed into the commercial entrepôts for trade with France. Just prior to independence, up to half Algeria's exports by value literally flowed through the wine exporters' *caves* in these ports to the French market (Isnard, 1975). Manufactured goods flowed in the opposite direction, as France considered its North African colonies as virtually a captive market. Traditional craft-based industries in renowned centres like Tlemcen, Fez and Tunis struggled to compete with mass-produced industrial products from France. Contingents of North African troops and conscripted workers served to stem France's manpower haemorrhage during the First World War, and were followed soon afterwards by voluntary labour emigration to France. A couple of unsavoury Twentieth Century colonial wars in Libya and in the Rif region of Morocco, 1921-26, served to shatter rural infrastructures and economies and to set in motion rural-urban migratory flows.

With independence came an intensification of earlier urbanisation tendencies (see Table 3.1). Whereas only 21 per cent of the French Maghreb's population was urban in 1936, this proportion had increased to 27 per cent or about 6.5 million people by 1956 when Tunisia and Morocco acceded to independence. By 1984, this had rapidly grown to 48 per cent and about 23 million town-dwellers. Tunisia is the most urbanised country with Tunis taking on the role of a primate city (Table 3.1). By 1980, its 1.1 million people represented 32.5 per cent of Tunisia's urban population (Troin *et al.*, 1985). Macrocephalic tendencies also typify the growth of Casablanca to 2.14 million people by 1982 and Algiers to 1.9 million by 1983. The second city of Algeria, Oran, used to contain 61 per cent of the population of Algiers in 1954. This ratio had decreased to 36 per cent in 1966 and 32 per cent in 1977; hence, concern about the threat of macrocephalic urban growth in Algeria (Mutin, 1986). Whereas Morocco does exhibit strong regional urbanisation focussed on Fez, Meknes and Marrakech, the coastal urban axis from Casablanca, through Rabat-Salé, the political capital, to Kenitra focusses 40 per cent of the country's urban population and 20 per cent of its total population. Several decades of continued rural-urban

Table 3.1: Recent urbanisation in the Maghreb

Census Date	Morocco Urban Population (millions)	Percentage Urbanised	Algeria Urban Population (millions)	Percentage Urbanised	Tunisia Urban Population (millions)	Percentage Urbanised	Libya Urban Population (millions)	Percentage Urbanised
1954	–	–	2.15	25.6	–	–	0.196	18.0
1956	3.4	28.4	–	–	1.25	33.0	–	–
1960	–	35.0	–	–	1.82	40.1	–	–
1966	5.4	–	3.7	31.5	–	–	0.39	25.0
1971	–	–	–	–	–	–	1.16	51.6
1973	–	–	–	–	2.67	47.9	–	–
1975	–	–	6.84	41.0	–	–	–	–
1977	–	–	–	–	3.325	52.5	–	–
1980	–	–	–	–	–	–	–	–
1982	8.73	43.0	–	46.0	–	54.0	–	61.0
1983	9.55	43.0	9.9	–	3.68	52.8	–	–
1984	–	–	–	–	–	–	–	–

Sources: Troin et al.(1985, p.150); World Bank (1985, p.216–7); Lawless and Kezeiri (1983, p.81–6)

migration lie behind this strong urbanisation. In the early 1950s, these three countries had to deal with about 70,000 rural-urban migrants a year. By the mid-1970s, this rural-urban exodus had grown to 300,000 people annually, which represented about 1.1 per cent of Morocco's rural population and about 1.4 per cent of Algeria and Tunisia's rural population departing each year to the expanding towns (Troin *et al.*, 1985). This compared with annual rates of 0.4 per cent in the 1940s.

Much of this rural-urban migration has been focussed on the larger metropolises especially those with significant recent industrial growth. The considerable role given to industrialisation within the economic development strategies of all Maghreb states, and especially those of Algeria and Libya, has served to further urban bias in terms of the frequent location of new industry in coastal nodes, often the old-established port cities. Whereas some raw materials-based industrialisation has prompted more peripheral regional development, as in Libya and Algeria's Saharan oil and gas fields (Schliephäke, 1977), the later processing industries based on these mineral resources have usually been at coastal and often urban locations. Morocco's growing phosphates-based chemical industry at Safi, Libya's petrochemicals complexes at Marsa el Brega and Ras Lanuf, and Algeria's iron and steel complex near Annaba and its petrochemicals complexes at Skikda and Arzew all exemplify this locational trend. Much other recent manufacturing expansion has reflected the market attractions of large established cities, not just the capital cities, but also other regional centres like Oran (Tidjane and Sutton, 1985) and Kenitra-Mohammedia (Kaioua, 1984). Not only do these new industrial employment poles attract further rural-urban migration, as much for their initial construction site jobs as for their eventual manufacturing employment, but they often set in motion considerable commuting systems on a daily and a weekly basis. The industrial zone of Rouiba-Reghaia and other developments in the eastern part of Greater Algiers generate commuting from much of the Mitidja Plain, the Grande Kabylie region, and even farther afield (Gaidon-Arvicus, 1980). Only more recently has emphasis been put on deconcentrating industrial development away from its continued spatial bias towards coastal urban locations.

By dint of its incompatibility with heavy industry, tourist development has tended to redress the regional development imbalance somewhat, especially in Tunisia and Morocco. Both countries have promoted foreign tourism as a major currency earner in their economic development strategies. Both attract 1.5-2 million tourists a year, largely to purpose-built resort complexes which, while away from the capital, do serve to urbanise their local regions. Thus the Hammamet, Sousse, and Djerba-Zarzis tourist poles in Tunisia now contribute to this urban-coastal bias. At least this can be regarded as a form of regional development, especially with related efforts at Saharan tourism in Southern Tunisia. In Morocco, Agadir is also a more Southern peripheral development focus though the Tangiers region and Marrakech tourist developments serve to foster established urban centres.

Elsewhere in the service sector post-independence expansion in higher education and health facilities initially favoured established urban locations (Table 3.2). Only in Algeria are second stage university centres being established in regional centres like Batna and Tlemcen, but generally accession to higher education or to specialised medical care facilities necessitates migration to the larger cities, often the capital. Even Tunisia's early promotion of a national family planning programme displayed a decidedly spatial bias. By 1975, 49.6 per cent of the women recorded as practising family planning resided in the three most urban administrative regions of Tunis, Sousse and Sfax. Similarly, 46 per cent of the doctors with family planning qualifications were concentrated in Tunis alone (Taamallah, 1978).

This pro-urban trend in Maghreb development since independence ought to have provided an expanding, increasingly affluent, and concentrated market for the region's agriculture. In the event, much of the demand-led impetus has flowed overseas as, in general, agriculturalists have failed to maximise their reaction to this stimulus and the Maghreb has increasingly become a food-deficit region. Costly imports of cereals rose from 3 million tonnes in 1974 to 5 million in 1979 and to 7-8 million in a poor year like 1981. Algeria's situation is particularly bad with, in the 1975-79 period, its dependency on imported cereals being 61

Table 3.2: Urban-rural contrasts in Algeria – 1984

Index	Urban	Rural
Percentage living less than 1km from a primary school	93.1	64.9
Percentage living within 5km from a secondary school	97.3	39.1
Percentage living within 5km from a maternity care centre	92.9	25.0
Percentage living more than 15km from a hospital	5.1	74.3
Percentage of total working population illiterate	27.9	55.3
Percentage of working population aged 25-29 illiterate	8.7	39.2
Percentage of industrial workers illiterate	25.2	43.1

Sources: Office National des Statistiques (1985a, p.2-3; 1985b, p.3-4)

per cent, oil and fats 90 per cent, and dairy products 50 per cent, and all forecast to be worse by 1990 (Troin *et al.*, 1985). While Morocco and Tunisia were respectively able to cover 69 per cent and 53 per cent of agricultural imports by agricultural exports in 1980, Algeria could only manage 6 per cent. Similarly in food deficit Libya, by 1984, domestic food production met only around 30 per cent of food needs. This food dependency also represents political dependency on EEC and North American cereal producers.

While institutional factors can be regarded as partly explaining this problem, particularly the failure to reform structures in Morocco and the inadequacy of the agrarian reforms in Algeria and Tunisia, perhaps deeper causes promote urban bias. The continued emphasis from colonial times on export crops like wine and citrus fruits, so as to provide foreign currency to finance urban-industrial development, served to neglect the possibilities of food crop production better geared to the domestic market. Only quite recently have Algeria's vineyards been converted (Sutton, 1987a) and the profitability of livestock production and market gardening come to be appreciated. This is largely because until the mid-1970s at least, producer prices and rural-urban terms of trade worked against agricultural producers and discouraged rural initiative and productivity. Lower consumer prices for basic foodstuffs prevailed as government policy aimed to keep urban and industrial wage levels low. This policy amounted to the transfer of the rural surplus to the towns. Thus, in Tunisia, producer prices for cereals dropped from an index of 100 in 1962 to 66 in 1970 and 55 in 1979. Terms of trade between rural and urban-industrial goods worsened and rising input costs for agriculturalists discouraged investment and intensification and so prompted stagnation. Rising urban demand through increased population numbers and their growing affluence therefore prompted further foodstuff imports during the 1970s, which themselves often had to be subsidised in the case of cereals, sugar and the like. From the mid-1970s, the increasing costs of both food imports and subsidies have prompted policy changes towards higher producer prices for the Maghreb's agriculturalists. But still in 1982, Algeria was spending 12 per cent of its annual budget on such price support for foodstuffs and Tunisia 15.5 per cent of its 1980 budget; both represent large sums of money, much of which theoretically could have gone to rural areas in the form of higher producer prices if the agricultural economy there were capable of producing more cereals, dairy products, and so on. Since 1975-76, production prices have been increased; in Algeria the index of cereal prices has passed from 100 in 1970 to 250 in 1981 (Troin *et al.*, 1985). By the mid-1980s, good harvests suggested that agriculturalists were making production adjustments to this new situation. But 'bread riots' in Tunisia and Morocco (Seddon, 1986) after urban foodstuff price rises have left the governments in a predicament, caught between high food subsidies and higher producer prices. Intensification is called for (Troin *et al.*, 1985) through irrigation, investment, and the general rehabilitation and up-grading of the Maghreb's rural sector.

Rural Development in an Urbanising and Industrialising Maghreb

This tendency towards 'urban bias' in post-independence development policies has not prevented the formulation of a range of rural, agrarian and agricultural programmes in each Maghreb country. Also, regional development programmes, in Algeria in particular, have been targeted on largely rural and peripheral regions, which have been long neglected and are still lagging in terms of comparative economic development. However, it can be demonstrated that the underlying 'urban bias' has served to deflect these ostensibly rural development actions away from the promotion of regional and peripheral self-reliant development and towards the further integration of such rural and peripheral regions into national urban-industrial development.

In this chapter, it is proposed to elaborate on this overall theme under three headings. Initially, the structural changes attempting to re-organise the Maghreb's agrarian systems will be reviewed. It can be argued that these have all been over-centralised. Secondly, the development of rural resources, especially water resources, can be seen as dominated by a large project mentality and the resulting irrigation schemes suffer from urban-industrial competition for the Maghreb's decidedly finite water resources. Thirdly, regional

75

development initiatives will be evaluated with respect to whether they are anything more than mere palliatives compared with the national plans which are usually dominated by urban-industrial development.

Structural Change and Agrarian Reform

In terms of the first theme of structural change, Algeria's agricultural system has probably undergone most reorganisation, starting with a 1960s changeover from French colonial farms to a smaller number of larger self-managed *autogestion* estates. Initially based on the spontaneous occupation of abandoned settler farms and later extended through the expropriation of remaining colonial land, this *autogestion* approach to 'agrarian reform' took over most of the modern agricultural sector of Algeria and reorganised it into about 3,000 self-managed units. These were rationalised in 1976 to 2,071 large units each averaging 1,164 hectares. A hierarchical elected structure, ranging from the General Assembly of Workers to the Management Committee and President, suggested self-managed autonomy, but the appointment of a Director by the agricultural ministry in Algiers centralised much of the decision-making. *Autogestion* workers became effectively agricultural labourers as the participatory and co-operative ideals were diverted, in part in pursuit of better productivity, in this vital commercial sector of Algeria's agriculture (Dominelli, 1974). After 1975, the centralised bureaucratic structures were partly dismantled, more local flexibility was granted, and the state marketing and supply agencies were replaced by more local and regional co-operative bodies.

The early 1980s heralded another shake-up of the *autogestion* sector (Figure 3.1). The 2,000 over-large estates were broken up into smaller units and often combined with production co-operatives produced by the subsequent 1970s agrarian reform undertaken in Algeria's private sector of agriculture. As a result, 3,415 *domaines agricoles socialistes* (DAS) were established, averaging 830 hectares in size (Cote, 1985). Though not much smaller than their precursors, it was intended that these DAS would be more specialised and, as a result of other training programmes, better staffed in terms of technicians, managers and accountants. An early

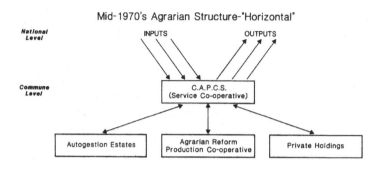

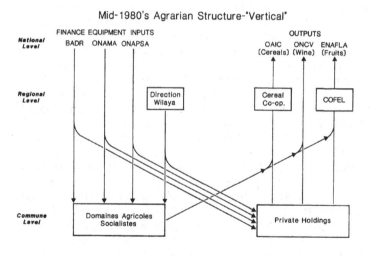

Figure 3.1: Algeria's agrarian structures in the mid-1970s and the mid-1980s (Source: Côte, 1985)

measure of success was the increasing number of DAS declared to be operating at a profit: 398 or 8 per cent in 1982-83 rising to 2,000 or 46 per cent of the total number in 1984-85 (Grand Maghreb, 1985). However, it has taken nearly two decades to start to achieve better production in this essentially state section of Algerian agriculture, during which period over-centralised *dirigiste* policies have largely prevailed.

To a certain extent similar organisational problems have beset Algeria's 1970s agrarian reform of its private and largely traditional agricultural sector. By the end of 1980, about 1.3 million hectares of communal and private land had been expropriated from large landowners and absenteeists, together with 1.1 million date palm trees and a few sheep. These were then redistributed to over 90,000 reform beneficiaries (Sutton, 1982a; Bedrani, 1981). These beneficiaries were grouped into several types of co-operatives of which the *Co-opératives agricoles de production de la Révolution Agraire* (CAPRA) were the most evolved. A higher tier of service cooperatives or CAPCS (*Co-opérative agricole polyvalente communale de service*) was also set up, theoretically to benefit other local private agriculturalists and *autogestion* estates as well as the reform co-operators. Amongst many criticisms of this agrarian reform (Abdi, 1975), the lack of genuine participation in decision-making by the so-called co-operators has emerged (Sutton, 1982b). Despite having a hierarchical democratic framework akin to the *autogestion* units, the production co-operatives failed to operate in a true co-operative manner in the face of state centralisation and technocratic pressures. As early as 1974, René Dumont commented that the production co-operatives were remaining in the hands of the administration and, furthermore, were from the outset over-mechanised with tractors and combine harvesters (Van Malder, 1975). As all reform recipients were obliged to join these production co-operatives, the essential voluntary principle of co-operation was lost. Given the basic individualism of the Algerian peasant and his unfamiliarity with, and mistrust of, co-operative agricultural work, it is unlikely that a simple land redistribution could have prompted a co-operative approach without the state's insistence on membership. Within the CAPRAs, decision-making proved to be less democratic

78

than was planned and an internal social hierarchy often emerged (Jazra, 1976; Le Coz, 1979). As well as failing to boost significantly agricultural production, the 1970s agrarian reform suffered from much avoidance of its land redistribution aims by both large and absentee landowners (Sutton, 1982a). The profitability of many of the CAPRAs was low and the co-operators continued to suffer low income levels. Consequently, in the early 1980s the agrarian reform sector was again restructured (Figure 3.1) with about 40 per cent of the reform land being added to the state sector's *domaines agricoles socialistes*. The other 60 per cent of the reform land ceased to be worked by co-operatives but reverted to a more individualistic private form of agriculture. Together this amounted to a rather contradictory restructuring which extended privatisation at the same time as increasing the state sector (Côte, 1985). So ended a decade of a 'top-down' imposed co-operative system. It is intended that a better organised system of *domaines agricoles socialistes* and a regenerated private sector will increase agricultural production especially through intensification as symbolised by recent *plasticulture* in the state sector and ventures like poultry-farming in the private sector.

A more short-lived restructuring took place in Tunisia's agricultural system (King, 1977). Following independence the Tunisian government sought both to recover European settlers' land and to modernise traditional agriculture by acquiring the modern well-equipped estates from departing Europeans and making them central to an amalgam of smaller holdings, the whole entity forming a new *unité co-opérative de production* (UCP). Then in 1964, the remaining European-held land, about 300,000 hectares, was nationalised. Thus by mid-1968, this agrarian reform or co-operative sector of Tunisian agriculture contained 478 UCPs, which grouped 900,000 people on about 805,000 hectares. Such statistics suggest severe over-staffing on the UCPs in contrast to the remaining 40 per cent of the agricultural land of Northern Tunisia, much of which belonged to just 3,000 large landowners practising more extensive forms of agriculture. Elsewhere in Northern Tunisia, 64,000 private smallholders practised intensive traditional agriculture on land often seriously degraded. In 1968, this agrarian reform with its supposedly 'co-operative' approach was extended into the Cap Bon and Sahel regions, both areas of small peasant proprietors who reacted against the prospect of co-operativisation. It was already realised that despite their elected councils, the UCPs were fairly centralised through their appointed directors who followed cultivation plans emanating from the Ministry of Agriculture. Political opposition extended from the Sahel's peasant proprietors through to the World Bank loan providers and together they managed to counter a Government move in 1969 to extend this agrarian reform rapidly to the rest of Tunisia's agricultural system. The attachment of the peasantry to smallholdings or patches of olive groves appeared greater than anticipated, the co-operative approach was poorly explained, and, moreover, the new agricultural system was imposed from the top. The result was the total failure and reversal of Tunisia's agrarian reform and most of the 4 million hectares of land which had briefly been reorganised into co-operatives was quickly dispersed or repossessed by the former owners. Land originally part of the European farms was placed with a national agency, the *Office des Terres Domaniales* (OTD) which organised some of it into pilot farms or UCPs but at least half of it was quickly rented out or sold to private cultivators. By 1977, the OTD was responsible for about 440,000 hectares, of which 232,913 hectares were still co-operativised within 228 UCPs (Kassab and Sethom, 1980). Thanks to better management and a reduction in overmanning, most of these co-operatives were registering profits during the mid-1970s. Elsewhere in the private sector, land concentration has resumed, as has urban absentee land ownership. The reversal of the agrarian reform experience of 1962-69 appears to have halted other progressive changes in Tunisian agriculture with social and urban-rural disparities reverting to their earlier high levels.

Landed interests, often urban-based or with urban links also served to reverse part of the 1970s Algerian agrarian reform discussed earlier, namely its so-called 'Third Pastoral Phase'. Launched in 1975, this concerned the reform and reorganisation of the pastoral sector wherein just 5 per cent of the 250,000 rearers and shepherds owned 50 per cent of the livestock and often exercised monopoly control over pasture. This Third Phase of Algeria's *Révolution Agraire* decreed that absenteeists, often urban, could no longer possess flocks, that flock sizes for 'direct' graziers would be limited, that excess and expropriated livestock would be redistributed to working shepherds and small graziers and that remaining shepherds would become proper wage-earners with full rights rather than 'share croppers'. A system of service co-operatives for pastoral activities would control pasture rights and seek to manage pastureland properly in the community's interest. Thus livestock would remain in private hands, albeit those of small and medium-sized graziers who could choose whether to form loose groupings or more communal co-operative organisations (CEPRA). A degree of sedentarisation in pastoral villages was also anticipated (MARA, 1976).

Despite these good intentions, progress with this 'Pastoral Phase' proved very slow. By 1978, only 12 out of 17 steppe region *wilayate* had even paper plans for CEPRA and only 45 such pastoral co-operatives were functioning, all in the *wilaya* of M'Sila. By late 1980, the programme had ground to a halt in the face of determined opposition from the large grazier interests who threatened to slaughter livestock or move them across frontiers to avoid expropriation. The Minister of Agriculture hoped to re-launch the Pastoral Phase of the Agrarian Reform in 1983 and to this end a 'High Commissariat for the Development of the Steppe' was set up. In the event, this imaginative if difficult pastoral reform faded from government policy along with the rest of Algeria's agrarian reform in the face of a policy re-evaluation in favour of a more privatised approach to agriculture.

An increasingly visible result of Algeria's agrarian reform is the scatter of supporting 'Socialist Villages' constructed to house those reform beneficiaries in need of better housing more accessible to their new reform land. By late 1981, 171 socialist villages had been completed and another 202 were under construction. By 1985, Brûlé and Fontaine (1986) estimate that about 400 socialist villages had been established, but by then the programme had been collapsed into a general rural housing policy detached from agrarian reform and more orientated towards private initiative. In time many criticisms emerged of the socialist villages programme (AARDES, 1978; Lesbet, 1983; Sutton, 1984 and

1987b). These particularly focussed on the lack of participation and involvement in designing and building the villages on the part of their prospective inhabitants and on the tendency for both house design and village layout to approximate to urban standards rather than making use of more appropriate rural settlement forms or building techniques. Both areas of criticism stem in large part from the over-centralised nature of the programme, which sought national standards and uniformity of approach through the imposition of a centre-down settlement policy. The self-evident needs of land reform recipients for improved housing and service centres were both local and rural, but the attempt to meet these requirements was directed by centralised decision-makers with largely urban values and perspectives. Thus dwellings were designed by technicians with limited knowledge of rural society. Inevitably they imparted their own cultural and urban values in house layout, the positioning of windows and doors, and the whole ordered morphology of the socialist village which gives the impression of a suburban housing estate rather than a functional agricultural village. Without local participation the construction of each village became a uniform and standardised operation which often failed to accommodate the local needs and regional idosyncracies involved in a host of diverse settlement situations. While house design may vary, the highly concentrated form of each settlement with straight, often quite wide streets and serried lines of houses follows a standardised package. Both house and settlement often confronted the new resident with a 'discontinuity of modernity', as indeed did the new co-operative mode of agricultural production. For some the adaptations necessary were too great and instances of desertion and apathy have been recorded. Womenfolk particularly suffered from the lack of privacy in the nucleated settlement and from unfamiliar and inappropriate house design. The rejection of vernacular rural house designs extended to building materials. Local and familiar stone or brick were displaced by concrete and breeze-block less suitable for prevailing climatic conditions and less easy to repair and modify. The concentrated peri-urban morphology of the socialist village especially has been contrasted with the loose clustering of traditional rural *douars* or hamlets and with the dispersed settlements of much of rural Algeria. It would appear that the technocratic urban builders projected their modern urban-industrial values on the peasantry in a novel form of economic and cultural domination which threatened the basic objectives of the agrarian reform, namely to improve rural levels of production and social well-being.

This mixture of over-centralised and urban-dominated approaches to structural change in the Maghreb's agrarian systems can be invoked as having perpetuated unequal rural-urban interaction or having worked against pro-rural redistribution of landed and other assets. The role of rural elites in this process has also to be acknowledged, as exemplified by Algeria's pastoral elite or by Tunisia's rural groundswell against agrarian reform and co-operativism. Whereas the rural elite may share with the rural poor an interest in more rural investment, they nevertheless share with the urban elite an interest in concentrating rural resources and in avoiding major redistributions of power and wealth. Lipton (1984) suggests that the resolution of this conflict of interest results in a political outcome in which rural elites receive concessions such as subsidies on inputs or credit or, as in Algeria, are allowed to by-pass land expropriation measures, in return for accepting urban-bias in output prices and in the allocation of human and physical capital. Such an alignment of rural elites with decision-making townsfolk can produce anti-rural, inefficient and inequitable outcomes. But then the rural elite is already urbanising its social and economic activity space through transferring residence, capital investments and offspring in education to the towns, while maintaining rural assets, electoral power and clients in the country. Further, the rural elite seeks to influence state policy for rural areas in ways that mainly help better-off rural groups, especially absentees seeking to transfer surpluses to urban investment outlets. The outbreak recently in Algeria of private house building in and around Algiers and other cities exemplifies the amount of private capital long present in the socio-economic system which has only been made visible following the 1980s Chadli Government's rehabilitation of the private capital sector previously marginalised by the over-dominant state companies approach to economic management.

Rural Water Resources

The second theme of rural development focusses on the technical development of rural resources, in particular water resources. In Morocco and, through force of circumstances rather than ideology in Libya too, water resource development has promoted large project irrigation schemes to the neglect of, and indeed as an alternative to, the kind of structural changes in agriculture discussed so far. Even in the limited area of the Maghreb where precipitation exceeds 400 mm a year, seasonal drought makes irrigation desirable; elsewhere it is a necessity. Hence, irrigation traditions are ancient, varying from oasis irrigation to the watering of mountain gardens and orchards. European colonisation re-introduced large scale irrigation works, neglected since Roman times. In Algeria, at least 17 large barrages were established by the French, largely in the West as in the Chéliff Valley. In Morocco, hydroelectric power production was an additional impetus to the development of large-scale irrigation barrages. After independence, both Morocco and Tunisia accelerated their programmes for irrigation development, while Algeria turned more to agrarian reform approaches and the alternative of industrialisation. The statistics in Table 3.3 illustrate the contrast existing by the early 1980s between Morocco's large-scale irrigation approach and Algeria's continued dominance by small and medium-sized projects.

In Morocco, large-scale irrigation barrages have become the symbol of the country's hoped-for *décollage économique* - development 'take off'. Thus, during the 1973-77 Plan, 1,586 million *dirhams* were allocated to new dam projects and to the main irrigated zones in comparison with 574 million allocated to the traditional cereal lands which represented 95 per cent of the cultivated area and employed most of the rural population. Further, given the singular lack of land reform within the irrigation perimeters, this investment served to benefit the already privileged minority of medium and large landowners. In the perimeter of the Barrage Youssef ben Tachfine (Tankist), 6 per cent of the proprietors owned 42 per cent of the land, compared with smallholders who composed 58 per cent of the proprietors and held just 12 per cent of the land. Similarly, in the

Table 3.3: Irrigated area in Morocco, Tunisia and Algeria

Country (date)	Total area irrigated (hectares)	Area of large scale irrigation (hectares)	Area of small & medium scale irrigation (hectares)	Percentage of cultivable area (hectares)
Morocco (1982)	786,000*	530,000*	256,000	11.0
Algeria (1981)	300,000	60,000	240,000	4.5
Tunisia (1981)	190,000	76,000	114,000	4.75

* of which 585,000 ha are equipped, 419,000 in the large scale sector

Source: Troin et al.(1985, p.90)

perimeter of the Oued Issen Barrage most of the benefits of intensification went to the 14 per cent of large landowners who together owned 70 per cent of the area (Revue de Presse, 1973). Meanwhile, Moroccan irrigation developments continue towards a target of one million irrigated hectares by 1990. Already by 1982, vast modern irrigated perimeters like the Tadla scheme (Figure 3.2) with 110,000 hectares irrigated and the Rharb with 86,000 hectares reflect progress in techniques and equipment and include industrial crops such as sugar beet, cotton, turnips and forage promoted by the *Office National d'Irrigation* (Troin *et al.*, 1985). By 1985, 32 dams had been built and another 34 were under construction or planned. However, this concentration of investment on large projects has been criticised (Findlay, 1984), in part on the grounds of inefficiency. Projects have been frustrated by the improper phasing between the provision of the dams and the development of the water distribution networks. Delays have retarded the return on these massive capital outlays. Despite the planners' assertion in 1973 that priority was to be given to completing existing projects, the 1978-80 Plan repeated the need to complete many of the same schemes. Findlay (1984) concludes that the official Moroccan policy of encouraging irrigation and favouring export-orientated crops in a few areas has had serious detrimental consequences for Morocco's wider agricultural system. Limited land redistribution has created an 'upper circuit' of agricultural activities linked to the international market but isolated from the bulk of the rural population and from the domestic demand for basic foodstuffs. Morocco's agricultural policies have thus accentuated the differences between the lower and upper circuits of its rural areas as well as enforcing a distinct pattern of spatial inequality in agricultural investment.

Furthermore, this large project approach has added considerably to Morocco's foreign debts. During the period 1981-1985, Morocco was to invest 4 billion *dirhams* in irrigation projects, in part financed from Gulf States and World Bank loans. Its dependency on these loan providers and on their willingness to reschedule some of these loans has increased, further tying the modern irrigated perimeters into external commercial systems and widening the gap

Figure 3.2: Irrigation developments in Morocco (Source: Troin *et al.*, 1985)

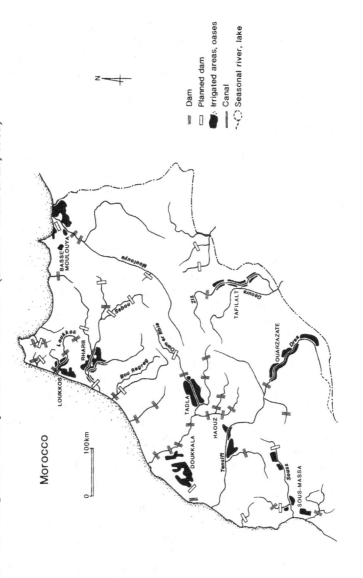

Dam

Planned dam

Irrigated areas, oases

Canal

Seasonal river, lake

between them and the more traditional rural production and settlement systems.

A similar large project approach has prevailed in Tunisia with its two Northern irrigation schemes, the Lower Medjerda Valley Project, initially conceived by the French, and the more recent Upper Medjerda Valley Scheme (Figure 3.3). These two are closely linked hydrologically but have always been organisationally quite separate. By 1981, the Lower Medjerda (31,300 hectares) and the Upper Valley (15,000 hectares) accounted for most of Northern Tunisia's 54,000 hectares of irrigable land. More recent large-scale developments in the Nebhana and Kairouan areas brought Central Tunisia's irrigable area up to 17,500 hectares. Planned developments for the 1981-86 period were to add another 60,800 hectares to the large-scale irrigation schemes, of which 39,800 would be in the North. Troin *et al.* (1985) doubts such a rate of expansion. It ought to be stressed that the major Lower Medjerda Valley Scheme has consistently fallen behind targets both in terms of area and the quality of crops produced. Competing urban-industrial demands for water from nearby Tunis have reduced the irrigable area and the use of poorer quality water has necessitated a reliance on more salt-tolerant and less valuable crops (Karabenick, 1969). In both Upper and Lower Valleys, the previously discussed abortive agrarian reform disrupted the development of irrigation. Thereafter, as in Morocco, investment in irrigation equipment had priority in the agricultural sector's investment plans, taking 44 per cent of the funding in the 1977-81 Plan. Within these plans underground water supplies are increasingly significant together with regional transference of water along a partly constructed North-South Canal linking the Medjerda Basin with the Cap Bon Peninsula and eventually to Sousse (Figure 3.3). Already, however, urban drinking water demands are literally diverting this irrigation canal towards the consumers of Greater Tunis (Revue de Presse, 1985).

Probably the ultimate in large-scale irrigation projects is to be found in Libya where practically all water comes from groundwater aquifers. Traditional well irrigation in the Jefara Plain near Tripoli is seriously depleting and salinising the coastal aquifer. This has enforced restrictions on more wasteful water-using crops and has even, ironically,

Figure 3.3: Irrigation developments in Tunisia (Source: Troin *et al.*, 1985)

prompted plans for agrarian reform to re-organise and plan more rationally Tripolitanian agriculture (Allan, 1982). The large-scale approach, however, has taken root farther south, deep in the Libyan Sahara at the Kufrah oasis. Started by the Occidental Oil Company in 1968, this groundwater scheme was taken over in 1972 by the government-owned Kufrah Agricultural Company. Based on fossil groundwater resources this project is divided into two schemes. Initially the Kufrah Production Project developed 10,000 hectares of fodder crops grown in circular 100 hectare plots, each irrigated from a central well from which water is distributed by an immense sprinkler. Initially this was to feed 250,000 sheep but later the scheme was turned over to cereal-growing instead. Secondly, there has been developed a less technocratic Kufrah Settlement Project consisting of 5,000 hectares of small farms, arranged interestingly in 55 hexagonal units, each of 100 hectares (Allan, 1975). Unfortunately, the first year or so of pumping resulted in a much more rapid "drawdown" of the aquifer than predicted and a major reassessment of the potential of the whole Kufrah project was necessary. Transport and managerial costs were appreciated as high, but now high environmental costs were added. A related project at Sarir oasis farther north has seen a wider spacing of the wells to lessen the over-rapid drawdown of water levels. Estimates in the late 1970s that the grain produced at Kufrah, once transported to Libya's coastal towns, costs ten to twenty times the world price for grain serves to condemn the whole technocratic project as a failure, a case of unlimited central government capital resources warping judgements as to what constitutes sustainable rural development. Urban foodstuff requirements appear to have prevailed over any consideration of what Libya's rural population needed in terms of development assistance. Nevertheless, the 1980s are seeing another ambitious water resources scheme here, namely the 'Great Man-made River Project' (Economist Intelligence Unit, 1984) which aims to transport immense quantities of Saharan underground water to the coastal agricultural areas. A 900 km pipeline under construction will carry two cubic km of water a year from Kufrah and Tazerbo to the coast and a later pipeline from Fezzan to Western Tripolitania will add another one cubic km of 'fossil water'. This will irrigate

250-300,000 hectares which could possibly result in food self-sufficiency until Libya's population, now about 3.8 million, exceeds 6 million. The costs of such a high technology solution should preclude the cultivation of low value crops like wheat. Fruit and vegetables for export should be grown with which revenue basic foodstuffs would be purchased on world markets. Again a decidedly technocratic urban commercial approach to rural development prevails, an approach which, as in Tunisia, may well be threatened in time by competing demands for the same fossil water from Libya's burgeoning towns and ambitious industrial developments.

Urban-industrial competition for rural water resources is thus becoming a general Maghreb development theme. Algeria's industrial and drinking water needs are growing rapidly, taking at least 20 per cent of water resources. Algiers's water consumption has more than doubled from 80 million cubic metres in 1969 and could attain 500 million by 2000. Already water collected for irrigation perimeters is being diverted to Algiers, Constantine (Figure 3.4), the Annaba steelworks, and elsewhere, despite investment in barrages which ought to triple the water available during the 1980s (Brûlé and Fontaine, 1986). Schemes are underway for quite considerable regional transference of water from the coastal Tell basins to interior High Plains cities like Constantine and Sétif (Côte, 1983). Thus, the present and future development of essentially rural resources, namely surface and underground water supplies, so as to promote more intensive agriculture, is compromised and subordinated to the interests of the urban-industrial sector. Urban-rural flows of capital and technological expertise are increasingly resulting in rural-urban flows of water.

Regional Development Policies

The rural peripheral parts of the Maghreb subject to these infrastructural and structural developments, often with an inherent urban bias, have also sometimes been the focus of regional development policies. Strong regional disparities, inherited from colonial times and often strengthened thereafter, have prompted calls for action. In the case of Algeria particularly, these have produced a range of regional

91

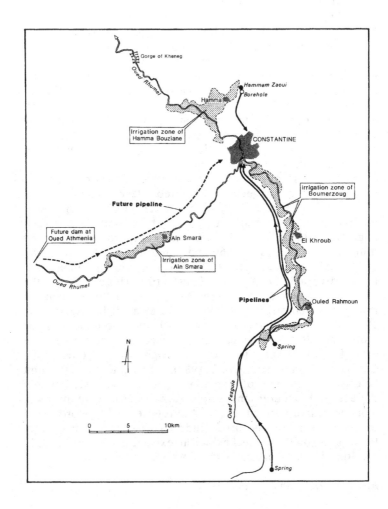

Figure 3.4: The diversion of irrigation water to Constantine (Source: Côte, 1983)

development policies to modify, though not to reverse, the centralised national development plans dominated by industrialisation and urban-orientated economic and social programmes. Given the imbalance in expenditure between national plans and urban-industrial private capital investment on the one hand, and regional development programmes on the other, the question can be posed as to whether such regional policies are anything more than a palliative in deference to certain regional and political pressure groups.

Algeria presents a more substantial catalogue of regional development initiatives than the other Maghreb countries. The heavy centralisation of decision-making was acknowledged by the Second Plan, 1974-77, which postulated six economic zones to cut across Algeria's established administrative divisions (Sutton, 1981). Each economic zone was allocated certain planning objectives which collectively aimed to lessen the growing concentration of population and economic activity in the coastal poles. However, these imaginative proposals were not fully accepted by the economic ministers and so were not put into effect, though some of the reasoning behind them was perpetuated in subsequent national plans. More successful were ten 'Special Programmes' launched intermittently between 1966 and 1973. Initially these were for complete *wilaya* or major administrative regions, but later covered only the lagging parts of the *wilaya* concerned (Nacer and Sutton, 1987). Compared with concurrent National Plans these regional programmes were quite substantial. The first four programmes amounted to 46 per cent of the 1967-69 'Pre-Plan' funding but spending then declined to 21 per cent of the First Plan, 1970-73. Spatially, the special programmes covered interior, Saharan and some coastal *wilayate* excluding the main urbanised regions. Sectorally, investments in housing, health and other social services amounted to over half the total expenditure. Agriculture accounted for more than a quarter of the funds, so contrasting with the industry-orientated national plans and suggesting that the aims were to transfer resources and social benefits to lagging regions. The special programmes can be criticised, however, for focusing too much of their funds on towns and for a lack of co-ordination with the National Plan's investments. But, they had introduced the spatial dimension into development planning and had emphasised the need for area-specific policies to deal with regional disparities. By the late 1970s the special programmes had been replaced by *Plans Communaux de Développement* (PCD). However, their total allocation represented only 6 per cent of the investments of the 1974-77 Second Plan, and only 31 per cent of these regional investments had been operationalised by the end of 1977. Given this declining ratio of regional as against national investment it is not surprising that only very limited regional convergence could be detected by Nacer (1979) for the period 1966-77. While the urban primacy of Algiers appeared to have declined, other coastal urban-industrial *wilayate* such as Oran and Annaba had prospered. Moreover, with respect to industrial employment indices, Nacer found continued regional divergence.

In view of this, the Algerian government attempted to diffuse development through enabling local authorities to promote local projects to develop their resources and meet some of their inhabitants' basic needs. Through this PCD system each of the 704 local communes could help elaborate and diffuse development. However, the annual expenditure only accounted for about 13 per cent of the national capital expenditure which, with its continued urban and industrial bias, gave rise to criticisms of regional development as a mere palliative. In National Plans, the main areas of investment are industry, economic infrastructure, housing, and education. By 1982 in the PCDs, hydraulics (domestic water supplies, sewerage, etc.) and social infrastructure formed the main sectors. In part, this is because industrial investment was covered by another centrally-administered programme, the *Petites et Moyennes Industries* programme (Nacer and Sutton, 1987). Within the PCDs framework a degree of urban bias prevailed with related *Plans de Modernisation Urbaine* for 39 selected towns to promote their urban infrastructure and service facilities.

Despite these regional development initiatives, the 1980-84 National Plan could still bemoan the concentration of population and economic activity in Northern Algeria and extrapolate a 1990 situation with 55-60 per cent of Algeria's burgeoning urban population in the five *wilayate* of Algiers, Oran, Blida, Constantine, and Annaba. In order to counter these trends, the 1980-84 Plan promoted action on an 'Options Hauts-Plateaux' to divert economic development into a broad belt of interior regions stretching east-west across eight *wilayate* to the south of the previously favoured Tell region. The Options Hauts Plateaux programme aims to add 1.5-2.0 million people to the region by 1990-95 and to increase its proportion of the national workforce from 22 to 27 per cent. Strong urban growth of 20 per cent per year would prevail in the region compared with 5.7 per cent nationally. Already work on upgrading the communications infrastructure has commenced with the first stage of a new railway line crossing the Hauts-Plateaux from Tébessa westwards to Saïda linking the main regional centres. Other provisions for the region include the building of four new towns, a car plant near Tiaret, and an aluminium works at M'Sila. This Options Hauts-Plateaux initiative was backed up by a better regional allocation of state investment funds, 1980-84. Table 3.4 demonstrates the improved share of funds planned to go to the region, 26 per cent compared with only 17 per cent during 1967-78. In the event, 28 per cent of a reduced actual expenditure supported this new regional development approach and the 'top five' *wilayates*' share declined dramatically. Given some flexibility and objective assessments of practical development possibilities this recent regional option represents a willingness in Algeria to break away from the traditional imbalanced spatial pattern of development (Nacer and Sutton, 1987; MPAT, 1980).

The subsequent Second Five Year Plan, 1985-89, endeavours to strengthen this regional planning option with a new *Projet de Loi sur l'Aménagement du Territoire* or 'Regional Planning Law' enacted in 1986. While recognising the planning problems involved in mastering urbanistion in Northern Algeria, this new 'law' prioritises the development of the Hauts-Plateaux region and the Saharan South of the country. New regional planning authorities, the *Schéma Régional d'Aménagement du Territoire*, (SRAT), promise to be an important innovation at a regional level, especially as administrative reorganisations in recent years have multiplied the number of *wilayate* so that a higher regional tier is now required. The planning region of each SRAT could well meet that need. Unfortunately, the mid-1980s drop in oil prices and hence of Algeria's national revenue has resulted in

Table 3.4: Regional allocation of public investment expenditure in Algeria

	1967-1978 percentage planned	1980-1984 percentage planned	1980-1984 percentage actual
Top 5 wilayate	38	23	18
South	16	8	8
Hauts-Plateaux	17	26	28
Rest of North	29	43	46
Total funds for Algeria (billion dinars)	252.2	167.68	75.47

Source: Nacer (1987)

widespread cuts in government spending, with regional development initiatives likely to bear the brunt of such reductions (Nacer, 1987). In both Tunisia and Morocco, less effort has gone into regional development policies despite strong regional disparities. Smallness of size and strong polarisation on Tunis as a primate city have partly accounted for Tunisia's reduced involvement, while successive Moroccan governments have shown less inclination to plan their economic development either nationally or regionally. Both during the planned economy approach of the 1960s and continuing with economic liberalism since, Tunisia has experienced a homogenisation of its national space with the whole country becoming the dependent region of Tunis (Troin *et al.*, 1985). Early regional links have largely dissolved to be replaced by national-scale interrelations articulated by Tunis, or indeed through Tunis to multinational company decision-making centres. The presence of an established urban hierarchy with regional centres like Sousse, Sfax and Bizerte has fostered this, especially as some of these plus Hammamet have become the poles of Tunisia's tourist industry which, while superficially spreading development, really serve to concentrate decision-making and capital controls in Tunis or indeed in Western Europe. Studies of development administration in Tunisia have confirmed this picture of centralisation and hierarchisation (Nellis, 1983). These found that decentralisation in general and even local administrative control of major development projects were not possible because of the lack of competent staff. A late 1970s USAID mission found that most regional *gouvernorat* authorities lacked one, two or even three of their seven key supporting staff and levels of competence and experience among those in post were low. From 1973, a limited Rural Development Programme has been set up but central officials retain close control of it. As many local officials were not elected locally but were rather appointees of a central ministry, they looked to Tunis for direction and so fostered centralisation. Discussion on decentralisation abounds especially in connection with the perennial speculation on post-Bourguiba Tunisia, but examples of regionalised policies are limited. One such example, launched in 1979, is the *Office de developpement de la Tunisie centrale* which on paper appeared innovatory, promising regional development for Central Tunisia. However, early results were poor and disorganised with ingrained habits of centralisation hard to overcome, not just within national ministries but also on the part of regional officials (Nellis, 1983). Prior to the mid-1970s, Morocco's spatial economy and administration remained highly centralised. Since then a degree of decentralised administrative reform has occurred and the country has been divided up into seven economic planning regions. Of these the Southern Economic Region, incorporating the disputed but *de facto* Moroccan Western Sahara has received a disproportionately large share of the funds dispensed by the Special Regional Development Fund, undoubtedly for strategic political reasons rather than for motives associated with genuine decentralised regional development (Nellis, 1983).

Of the Maghreb's peripheral regions, the greatest challenge for regional development comes from the Saharan South. The diagram reproduced here as Figure 3.5, originally offered by Côte (1983), demonstrates the break-up of the Algerian Sahara's traditional exchanges and interactions, these being replaced by outflows of oil and gas as well as Myrdal-type backwash flows of skills and capital to the development poles of Northern Algeria. Caravan trade and nomadic pastoralism have declined in particular, without which oasis agriculture is threatened with economic and social imbalance (Wilkinson, 1978). Regional development programmes in the former Oasis and Saoura *wilayate* have improved social infrastructure and revealed additional mineral assets but, as Figure 3.5 postulates, these too could flow out for the benefit of the urban-industrial poles rather than the Saharan periphery. Isolation, inaccessibility and negative social images combine to make the Sahara a social-economic desert as well as an environmental one.

With the exception of Algeria's succession of policies, regional development and decentralisation of administrative decision-making have made only limited progress in the Maghreb and, as a consequence, regional disparities remain strong, and may be strengthening. The extent to which this has resulted from intransigent urban elites is debatable (Nellis, 1983). Certainly established urban, and especially

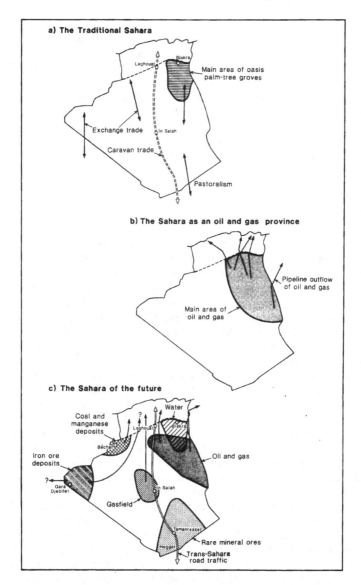

Figure 3.5: The Algerian Sahara: flows and interactions (Source: Côte, 1983)

capital city interests, have had an influence, but so too have constraints associated with the weakness of regional infrastructures and administrations and with the contradictory nature of many ostensibly regional development policies. Thus improved educational facilities can raise aspiration levels amongst school-leavers which can only be satisfied through rural-urban migration to better paid jobs or higher education centres. Thus regional development and associated decentralisation are slow painstaking ventures, providing plenty of evidence for the critics to label such endeavours as cosmetic and diversionary.

Conclusions

Since independence the Maghreb countries have all displayed increasingly centralised national space economies often with growing regional disequilibria. Some recent attempts at rural and regional development have as yet had little impact on this pattern of unequal development. This would come as no surprise to writers on development such as Mountjoy (1982), Taylor (1979) and Friedmann and Weaver (1979) who have observed, elsewhere than the Maghreb, that the neo-classical growth paradigm dominates theory and practice in many developing countries. Such conventional economic development remains largely based on promoting urban-industrial growth centres or axes, like Oran-Arzew or Casablanca-Kenitra, from which growth as modernisation should diffuse to the rest of the spatial economy. Despite criticisms by Stöhr and Tödtling (1977), Stöhr and Taylor (1981) and others, governments like those of the Maghreb remain largely wedded to centre-down development policies seeking greater functional integration of their national space so as to improve the spread of modernisation, industrial development and social infrastructure. Development is seen to filter through the urban hierarchy as part of an inexorable, if slow, march towards balanced growth. This belief prevails despite evidence that functional integration in the Maghreb space economy is leading to urban congestion in Algiers and Tunis, to regional inequalities between the Algerian Tell and the interior or between Northern and Southern Tunisia, and to continued rural poverty exacerbated by the lack of agrarian reform in Morocco and, indeed, despite limited agrarian reform in Algeria. Centre-down strategies prevailed, strengthened by the Williamson hypothesis that regional disparities will eventually lessen after having initially worsened as polarised rural-urban flows dominate in the early stages of change. Only partial modification was provided by Rondinelli's (1983) thesis offering spatial integration from below, with 'secondary cities' rather than primate cities polarising economic growth.

In the development literature, though not particularly in the Maghreb's development practice, the 'development from below' school (Stöhr and Taylor, 1981) gained strength despite limited practical case-study evidence. Centre-down strategies and the 'geography of modernisation' were criticised as ethno-centric, urban-biased, and too orientated towards the macroscale, whereas "the real problems were at the microlevel, with rural populations, with man-environment relations, with the grass roots" (Soja, 1979, p.34). It was increasingly argued that "the predominant process affecting the Third World was thus not so much the diffusion of development as it was the diffusion of underdevelopment" (Soja, 1979, p.34). Core-periphery structures at international and local levels had developed to maintain continuing conditions of dependency. The Maghreb offers a pertinent example with its Saharan hinterland increasingly capable of interpretation as a marginalised periphery (Abdi, 1982). Elsewhere in Africa, work on Kenya by Soja initially prompted his 'geography of modernisation' ideas and then later led to his self-criticism and revision of that paradigm (Soja, 1979; Taylor, 1979). Initially, Soja considered that improved communications increased urbanisation and the growth of rural-urban interaction prompted the diffusion of an exchange economy and of related modernisation including educational development and participation in national level political parties and trade unions. However, whereas Kenya displayed strong economic growth it was accompanied by widening regional income disparities and growing social disparities within the periphery. Hence Soja's attention shifted to the analysis of the social and spatial forces which created and sustained this disequilibrium, represented also in the Maghreb in increasing regional disparities, primate cities bursting with people but not with jobs, persistent rural poverty, and unequal terms of trade. The urban informal economy appears to be seen by many as offering better prospects than landlessness and unemployment in the rural sector. Hence excessive rural-urban migration continues. Other important processes are kinship-based economic remittances from the Maghreb's towns and from France which, with associated return and circular migration (Lawless, 1986), act as important redistributive and spread mechanisms, without which rural-urban disparities would be even greater. Taylor (1979) advocates more bottom-up strategies which would give primacy to the needs of the informal sector in both urban and rural areas rather than to the modernised formal sector. More attention should be focused on peripheral rural sub-systems, especially the informal sector of those sub-systems, that is attention to the development of small towns and their associated rural hinterlands. Such strategies would emphasise the beneficial linkages currently existing between the rural and urban informal sectors with a substantial reduction of the present dominance of the urban-controlled formal sectors. Some beginnings of such a bottom-up or periphery-inwards approach were detected by Sutton (1981) for Algeria, largely based on the decision-making structures and approaches of the Agrarian Revolution in the early and mid-1970s. But by the early 1980s, much of that promised decentralisation had evaporated along with the basic agrarian reform structures, leaving the decidedly more centre-down deconcentration approach with its regional development and regional industrialisation mechanisms. In short, urban bias had prevailed.

Whereas the superficial regional disparities and spatial patterns of urban bias can readily be described in the Maghreb, as exemplified by Côte's (1983) innovative study of Algeria, the mechanisms are less clear especially when partly masked by other spatial cleavages associated with ethnicity, class, and regionalism. Riddell's (1985) analysis of the various ways in which urban bias operates in Sierra Leone offers insights into similar mechanisms in the Maghreb. The mechanisms of urban bias include:

1. National spending and allocation priorities, especially 'development projects'. The dominance of industry in successive national plans, as in Algeria and particularly involving heavy industry in growth pole situations, exemplifies this. Morocco's phosphatic chemical industries and Libya's hydrocarbons projects can also be cited.

2. Pricing and taxing policies of state agencies like marketing boards. Figure 3.1 illustrates the ways purchasing boards like ENAFLA for fruit and ONCV for wine have controlled Algerian agricultural producer prices until recent relaxations have permitted more direct private sales of produce.

3. The provision and location in urban centres of social services and public utilities. Higher education facilities and health service infrastructure illustrates this facet of urban bias (Table 3.2).

4. The setting of 'cheap' urban food prices, especially for basics. While certain foodstuffs are expensive or even unobtainable in Algeria, bread, couscous, sugar and the like are inexpensive and subsidised, even if imported. To try and raise these prices for basics can produce urban unrest and riots as in Tunisia in early 1984.

5. The altering of traditional communal land tenure systems to allow for land agglomeration and increased production for the market. While more particular to Riddell's study area of Sierra Leone, both Algeria and Tunisia's agrarian reforms converted communal *arch* land and religious *habous* land into 'private' land, albeit then organised into a co-operative production system.

6. The location and protection of infant industries. Apart from late 1970s attempts at deconcentration for light industries around Algeria, an urban focus has prevailed for most industrialisation in the Maghreb (Mutin, 1985).

7. ".... a philosophy which informs national decision-makers that development is best pursued by a leading sector,

103

growth-pole, urban-industrial strategy" (Riddell, 1985, p.375). Algeria's hydrocarbons-based 'industrialising industries' strategy fits this facet of urban bias nearly perfectly.

Within these mechanisms Riddell identifies a triple squeeze on the peasantry stemming from export crop production, urban food provision, and labour force supply. A similar process can be discerned with regard to the Maghreb's agriculture. The urban poor's subsidised food in part results from policies which squeeze peasant livelihoods. This can be furthered by an agrarian reform attack on communal land, which through an extension of individual ownership of land can result in land agglomeration as well as lessening local propensity for the selective closure of rural economies. A further twist to Riddell's urban bias interpretation of agrarian reform is exemplified by Algeria's 1970s Agrarian Revolution which, Roberts (1984) has argued, had in part the objective of enthusing younger elements of the moribund and largely urban FLN party with fresh revolutionary zeal for the reorganisation of the country's agricultural system. As such, the agrarian reform served as a rallying point to politicise and radicalise the younger elements of the FLN party, which had increasingly come to be dominated by conservative cliques.

The above mechanisms of urban bias are able to operate because the urban and technocratic elite, so often in the Maghreb dominant within the apparatus of the state, are organised and literate (Lazreg, 1976). By contrast, rural and peasant interests are widely dispersed, poor, unorganised, and inarticulate. Moreover, many political and family links between town and country promote urban bias. Large landowner and grazier interests in Tunisia and Algeria were thus able to influence and deflect agrarian reforms which threatened to redistribute their assets to more genuinely rural agriculturalists.

Hence, it can be argued that the state plays a role in this perpetuation of urban bias, albeit sometimes transformed and modified by a mixture of cosmetic and genuine if deflected policies of rural and regional development. Because of their continued basic belief in urban-industrial led growth, the Maghreb governments have tended to redistribute resources to urban areas and hence to the enhancement of the state's own personnel in terms of their employment, remuneration, and status. Present trends in the Maghreb countries suggest that there is unlikely to be a major reversal of this urban bias. At most, there may be some promotion of secondary city-based regional development along the lines of Algeria's *Options Hauts-Plateaux*. Elsewhere, primate city-led regional disequilibrium and its associated flows of capital, people, and power is likely to persist in the Maghreb. Even in Algeria the prospect of a Greater Algiers in excess of 3 million people set within a more rapidly developing Tell region threatens to swamp strategies aimed at the promotion of a counter-balancing interior belt of urban-industrial centres in the High Plains. An increasingly urbanised Maghreb with ever more centralised space economies remains in danger of intensifying both the urban bias in its development strategies and the regional disequilibria which result from them.

Rural-Urban Disparities in the Maghreb

REFERENCES

AARDES (1978) *Les premiers villages socialistes - situation actuelle et perspectives*, Association Algérienne pour la Recherche Démographique, Economique et Sociale, Algiers

Abdi, N. (1975) 'La réforme agraire en Algérie', *Maghreb-Machrek* 69, 33-41

Abdi, N. (1982) 'Common regional policy for Algeria and Libya. From Maghrebi unity to Saharan integration', in: Joffe, E.G.H. and K.S. McLachlan, *Social and Economic Development of Libya*, Menas Press, London, 215-31

Allan, J.A. (1975) 'The hexagon is alive and reproducing in the Sahara', *Area* 7, 275-78

Allan, J.A. (1982) 'Capital has not substituted for water in agriculture', in: Allan, J.A. (ed.), *Libya Since Independence*, Croom Helm, London, 25-36

Bedrani, S. (1981) *L'agriculture algérienne depuis 1966. Etatisation ou Privatisation*, Office des Publications Universitaires, Algiers

Brûlé, J.C. and J. Fontaine (1986) *L'Algérie. Volontarisme étatique et aménagement du territoire*, URBAMA, Tours

Côte, M. (1983) *L'Espace algérien*, O.P.U., Algiers

Côte, M. (1985) 'Campagnes algériennes: un héritage colonial difficile à assumer', *Méditerranée* 55, 41-50

Dominelli, L. (1974) 'Autogestion in Boufarik', *Sociologia Ruralis* 14, 243-60

Economist Intelligence Unit (1984) *Quarterly Economic Review. Libya, Annual Supplement*, E.I.U., London

Findlay, A. (1984) 'The Moroccan Economy in the 1970s', in: Lawless, R. and A. Findlay (eds.), *North Africa Contemporary Politics and Economic Development*, Croom Helm, London, 191-216

Friedmann, J. and C. Weaver (1979) *Territory and Function: the evolution of regional planning*, Edward Arnold, London

Gaidon-Arvicus, A. (1980) 'Industrialisation et organisation spatiale dans l'Algérois. Leurs incidences en Mitidja orientale', *Cahiers de l'Aménagement de l'Espace* 10-12, 1-373

Grand Maghreb (1985) 'Economie et Société', *Grand Maghreb* 45, 476-477

Isnard, H. (1975) 'La viticulture algérienne. Colonisation et décolonisation', *Méditerranée* 23(4), 3-10

Jazra, N. (1976) 'Révolution Agraire et Organisation de la Production', *Terre et Progrès* 5(10), 5-12

Kaioua, A. (1984) 'L'Espace industriel atlantique marocaine de Kénitra à Mohammedia (Fascicule 13 de la Collection URBAMA), URBAMA, Tours

Karabenick, E. (1969) 'Irrigation Development in Tunisia', *Tijdschrift voor Economische en Sociale Geografie* 60, 360-68

Kassab, A. and H. Sethom (1980) *Géographie de la Tunisie. Le pays et les hommes*, Université de Tunis, Tunis

King, R. (1977) *Land Reform: a world survey*, Bell, London

Lawless, R. (1986) 'Return migration to Algeria: the impact of state intervention', in: King, R. (ed.), *Return Migration and Regional Economic Problems*, Croom Helm, London, 213-242

Lawless, R. and Kezeiri, S. (1983) 'Spatial aspects of population change in Libya', *Méditerranée* 47, 81-86

Lazreg, M. (1976) *The Emergence of Classes in Algeria*, Westview Press, Boulder

Le Coz, J. (1979) 'Dynamique de la Révolution agraire algérienne. La phase de la "bataille de gestion"', *Méditerranée* 35, 93-97

Lesbet, D. (1983) *Les 1000 villages socialistes en Algérie*, Syros and O.P.U., Paris and Algiers

Lipton, M. (1984) 'Urban bias revisited', *Journal of Development Studies* 20, 139-66

MARA (1976) *Textes relatifs à la troisième phase d'application de la Révolution Agraire*, Ministère de l'Agriculture et de la Réforme Agraire, Algiers

Mountjoy, A.B. (1982) *Industrialisation and Developing Countries* (5th edition), Hutchinson, London

MPAT (1980) *Rapport Général. 1er. Plan Quinquennial 1980-1984*, Ministère de la Planification et d'Aménagement du Territoire, Algiers

106

Mutin, G. (1985) 'Industrialisation et urbanisation en Algérie, in: URBAMA, *Citadins, villes, urbanisation dans le monde Arabe aujourd'hui* 87-113, URBAMA, Tours

Mutin, G. (1986) 'Aménagement et développement d'Alger', *Bulletin de la Société Languedocienne de Géographie* 20, 299-318

Nacer, M. (1979) *Regional Disparities and Development in Algeria*, M.A. Thesis, Department of Town and Regional Planning, University of Sheffield

Nacer, M. (1987) 'Evolution of Regional Development Policies in Algeria: a review paper', unpublished paper, Graduate School, London School of Economics

Nacer, M. and K. Sutton (1987) 'Regional disparities and regional development in Algeria', in: Forbes, D. and N. Thrift (eds.), *The Socialist Third World*, Basil Blackwell, Oxford, 129-68

Nellis, J.R. (1983) 'Decentralisation in North Africa: problems of policy implementation', in: Cheema, G.S. and D.A. Rondinelli (eds.), *Decentralisation and Development*, Sage Publications, Beverly Hills

Office National des Statistiques, Algeria (1985a) 'Les distances entre les residences des ménages algeriens et les infrastructure sanitaires et educatives', *Données Statistiques* 20, 2-3

Office National des Statistiques, Algeria (1985b) 'Analphabétisme 1984', *Données Statistiques* 21, 3-4

Revue de Presse (1973) 'La Politique des Barrages' *Revue de Presse*, No.178 (Algiers)

Revue de Presse (1985) 'Pour un dialogue et une co-opération sud-sud', *Revue de Presse*, No.299 (Algiers)

Riddell, J.S. (1985) 'Urban bias in underdevelopment: appropriation from the countryside in post-colonial Sierra Leone', *Tijdschrift voor Economische en Sociale Geographie* 76, 374-383

Roberts, H. (1984) 'The politics of Algerian Socialism' in: Lawless, R. and A. Findlay (eds.), *North Africa. Contemporary Politics and Economic Development*, Croom Helm, London, 5-49

Rondinelli, D.A. (1983) *Secondary Cities in Developing Countries*, Sage, London

Schliephäke, K. (1977) *Oil and Regional Development. Examples from Algeria and Tunisia*, Praeger, New York

Seddon, D. (1986) 'Bread riots in North Africa: economic policy and social unrest in Tunisia and Morocco, in: Lawrence, P. (ed.), *World Recession and the Food Crisis in Africa*, James Currey, London, 177-192

Soja, E.W. (1979) 'The geography of modernization - a radical reappraisal', in: Obudho, R.A. and D.R.F. Taylor (eds.), *The Spatial Structure of Development: a study of Kenya*, Westview, Boulder, 28-45

Stöhr, W.B. and D.R.F. Taylor (1981) *Development from Above or Below?*, Wiley, Chichester

Stöhr, W.B. and F. Tödtling (1977) 'Spatial equity - some anti-theses to current regional development doctrine, *Papers of the Regional Science Association* 38, 33-53

Sutton, K. (1981) 'Algeria: centre-down development, state capitalism and emergent decentralization', in: Stöhr, W.B. and D.R.F. Taylor (eds.), *Development from Above or Below?*, Wiley, Chichester, 351-375

Sutton, K. (1982a) ' Agrarian reform in Algeria - progress in the face of disappointment, dilution, and diversion', in: Jones, S., Joshi, P.C. and M. Murmis (eds.), *Rural Poverty and Agrarian Reform*, Allied Publishers, New Delhi, 356-75

Sutton, K. (1982b) 'Agricultural co-operatives in Algeria', *Yearbook of Agricultural Co-operation 1981*, 169-90

Sutton, K. (1984) 'Algeria's socialist villages - a reassessment', *Journal of Modern African Studies* 22, 223-48

Sutton, K. (1987a) 'Algeria's vineyards - a problem of decolonisation', *Méditerranée*, (in press)

Sutton, K. (1987b) 'The socialist villages of Algeria' in: Lawless, R.I. (ed.), *Middle East Villages*, Croom Helm, London, 77-114

Rural-Urban Disparities in the Maghreb

Taamallah, K. (1978) 'La régulation des naissances en Tunisie', *Population* 33(1), 194-205

Taylor, D.R.F. (1979) 'Spatial aspects of the development process', in: Obudho, R.A. and D.R.F. Taylor (eds.), *The Spatial Structure of Development: a study of Kenya*, Westview, Boulder, 1-27

Tidjane, B. and K. Sutton (1985) 'The spatial structure of state-owned industry in the Oran region of Algeria', *Tijdschrift voor Economische en Sociale Geografie* 76, 261-73

Troin, J.F. *et al.* (1985) *Le Maghreb. Hommes et Espaces*, Armand Colin, Paris

Van Malder, R. (1975) 'La révolution agraire en Algérie: tournant politique ou infléchissement technique?', *Civilisations* 25, 251-71

Wilkinson, J.C. (1978) 'Problems of Oasis Development', *Research Paper 20, School of Geography, University of Oxford*

World Bank (1985) *World Development Report 1985*, Oxford University Press, Oxford

4

THE URBAN CRISIS AND RURAL-URBAN MIGRATION IN SUDAN

EL-SAYED EL-BUSHRA

Introduction

In this chapter the status, causes and consequences of urbanisation and urban-rural interaction in Sudan will be discussed. In the first part, an account of urban development in Sudan is given, reflecting socio-economic changes which have been taking place since the turn of the century. The second part of the chapter concentrates on the causes and implications of internal and international migrations. The impact of migration on Sudan's urban system and the problems pertaining to urban bias and the degree of interaction between urban and rural areas will be stressed. Special attention will also be given to the question of migration from Sudan's peripheral or marginal areas to the riverine core which is epitomised by the Khartoum conurbation - better known as the triple capital - composed of the cities of Khartoum, Khartoum North and Omdurman, located at the confluence of the Blue and White Niles.

Important contributions to the study of urbanisation and urban development in Sudan have been made by various specialists, including scholars from geography, sociology, demography, history, architecture and planning. But although the literature dealing with the topic is extensive, too little light has been shed on the positive and negative aspects of urbanisation. Most of the writings have tackled the general issues of urbanisation and urban development (Hamdan, 1960; El-Bushra, 1967-1985/86; Hale, 1971; Lobban, 1975; Farghali, 1977; Winters, 1977; Pons, 1980; El-Arifi, 1980; Nour, 1986; El-Sammani et al., 1986). Fewer contributions have concentrated on the specific problems associated with urbanisation, including those related to rural-urban migration (Henin, 1962; Galal-al-Din, 1980; Hijazi, 1981; Herbert and Hijazi, 1984; Al-Agraa *et al.*, 1985). In this chapter, an attempt will be made, therefore, to highlight the positive and negative aspects of rural-urban interaction in Sudan, with particular reference to the Khartoum conurbation. It is the contention of the present writer that negative aspects by far outweigh positive ones. Heavy migration from rural to urban areas, particularly as most of the migration is directed towards the capital city, is bound to have serious repercussions for both the rural and urban areas. While the process of rural-urban migration deprives the rural areas of their most skilful and productive elements, the primate city is overwhelmed by migrant labour for whom there are limited opportunities for employment. To complicate the situation still further, the emigration of large numbers of professionals and artisans from Sudan to the oil-rich countries of the Middle East adds a new dimension to the internal development problems of the country (Weaver, 1985; Davies, 1986; Lewis, 1986). Moreover, the armed rebellion which has broken out intermittently in Southern Sudan since independence in 1956, the civil wars and political instability in neighbouring countries, together with the harsh years of drought and famine which have devastated the African Sahel, have all contributed to the heavy waves of migration into Central Sudan. Sudan takes most of the burden of the refugee problem of Africa by receiving large numbers from neighbouring countries such as Ethiopia, Uganda and Chad (Weaver, 1985). Most of the migrants coming either from internal or external sources have headed towards the principal cities. Under the prevailing conditions, the whole phenomenon of urban-rural interaction will be thrown off-balance. Indeed, it can be suggested that the internal security and political stability of the country at large is put into real jeopardy because of this mass movement of immigrants.

Finally, in addition to secondary sources, the data used in this study are drawn mainly from the material provided by the 1955/56, 1973 and 1983 population censuses, together with information furnished by the 1964/65 population and housing survey. The chapter also draw on material collected by means of a limited questionnaire survey which was conducted by the author in May 1987, with a group of migrants representing the various areas of out-migration in Sudan.

Urban Development in Sudan

Northern Sudan is one of the few areas in Tropical Africa which have a long urban tradition. Indeed, urban life in this part of Sudan has existed since the prehistoric era (El-Bushra, 1971a, 1971b, 1972a, 1975c; Winters, 1977). Throughout history, urbanised communities have developed to perform administrative, commercial and educational tasks. However, historic urbanisation was both limited in scope and its spatial distribution. Modern urban development commenced with the colonial rule at the turn of this century. The socio-economic changes which were introduced during the colonial period 1898-1956 necessitated the establishment of new towns to perform the role of administrative and other functions (El-Bushra, 1969). The recruitment of rural people to fill the newly created jobs in towns marked the beginning of rural-urban migration on any sizeable scale. In order to facilitate the government of the country, the colonial rulers resorted to the establishment of administrative centres. Moreover, a modern network of transportation together with commercialised farming were introduced for the first time to boost export trade. These economic transformations stimulated the development of new towns to act as nodes of communication and centres of trade. The domination of political life by the urban elite since independence has placed even more emphasis on the role of cities in the general process of social and economic development, at the expense of the rural areas. At present, however, the rationale behind the process of rural-urban migration takes a completely different orientation. Whether it is drought or famine, crop failure or disguised unemployment, civil war or lack of basic amenities, migration seems to be the only option open for the rural poor (Gilbert and Kleinpenning, 1986). Unfortunately, the economic base of cities, in an underdeveloped country like Sudan, is such that these centres are unable to cope with the

constant flow of large numbers of migrants (Figure 4.1).
The results in the recipient cities are catastrophic, leading to
a serious deterioration in the urban environment, as
expressed by the incidence of squatter settlements around the
Khartoum conurbation (Figure 4.2).

The first statistics to reveal the level of contemporary
urbanisation in the Sudan were released by the 1955/56
Census, wherein only 8 per cent of the Sudanese were
classified as urban. According to the 1983 Census, 20 per
cent of the total population was considered urban, indicating
that Sudan is still only slightly urbanised, even by African
standards. The projection for the year 2000 also indicates a
low level of urbanisation (Tables 4.1 and 4.2). However,
Sudanese urbanisation is highly polarised, and the shift from
rural to urban areas has been taking place at remarkable
speed in recent years (El-Bushra, 1974/75, 1975c, 1976a; El-
Arifi, 1980; Hijazi, 1981; Nour, 1986). Most of the major
cities are growing at an annual rate of 5-10 per cent. The
annual rate of increase for the Khartoum Conurbation
between the 1973 and 1983 Censuses was put at 8.5 per cent,
of which 5.6 per cent was attributed to rural-urban
migration (El-Bushra, 1979; Nour, 1986). The volume of
migration to Khartoum from the regions of Sudan in 1973 is
witnessed by Figure 4.1. The rate of growth due to
migration has increased tremendously during the period
1983-1987 because of the spread of drought and famine and
the intensification of military activities in the southern
region. Accordingly to the 1983 Census, the ten largest
cities accommodated 55 per cent of Sudan's urban dwellers,
indicating the sort of urban primacy which is typical of
many countries in the Third World (El-Arifi, 1980; Connel
and Curtain, 1982). In the case of Sudan, the Khartoum
conurbation has continued since its inception in 1900 to be
the unrivalled primate city (Figure 4.3). Between the 1983
Census and 1987, the population of the urban complex as a
whole is estimated to have grown from 1.34 million to 3
million, amounting to an increase of 123 per cent during this
five-year period. In fact, between 1900 and 1987, the
population of the triple capital multiplied sixty times from
50,000 to 3,000,000. This phenomenal growth has been
mainly attributed to rural-urban migration generated by
environmental, economic and political forces. In the sample

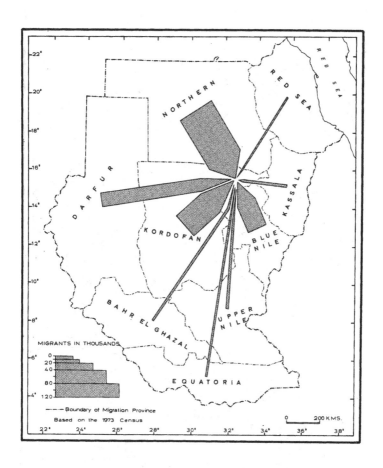

Figure 4.1: Migration to the Khartoum Conurbation, 1973

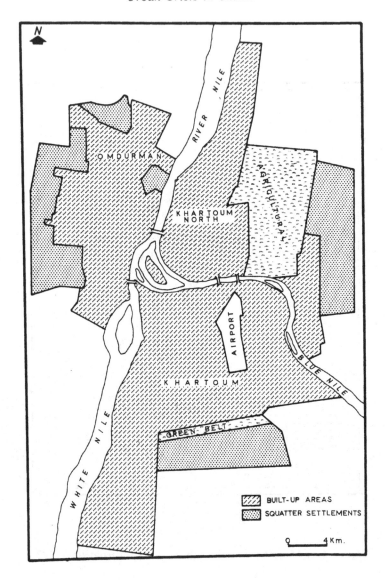

Figure 4.2: Squatter settlements in Khartoum, 1985

Table 4.1: Sudan – modes of living by region (1983 Census)

Region	Urban		Rural Sedentary		Rural Nomadic		Total	
	Number	% Total population	Number	% Total population	Number	% Total population	Number	% Total population
Khartoum	1,343,651	75	370,648	20	88,000	5	1,802,299	100
Eastern	638,833	29	1,010,700	46	558,676	25	2,208,209	100
Northern	230,341	21	802,414	74	50,269	5	1,083,024	100
Central	825,064	21	2,943,246	73	244,233	6	4,012,543	100
Kordofan	388,539	13	1,923,716	62	781,039	25	3,093,294	100
Equatoria	176,544	13	1,229,637	87	–	–	1,406,181	100
Darfur	316,152	10	2,307,803	75	469,744	15	3,093,699	100
Bahr el Ghazal	181,925	8	2,083,585	92	–	–	2,265,510	100
Upper Nile	52,510	3	1,547,090	97	–	–	1,599,605	100
Sudan	4,153,559	20	14,218,844	69	2,191,611	11	20,564,364	100

Source: derived from Unpublished Reports, 1983 Census

Table 4.2: Sudan - proportion of population urban and
urban-rural ratio, 1955-2000

Year	Percentage of population urban	Urban-rural ratio
1955	8	0.09
1965	12	0.14
1975	16	0.20
1985	20	0.25
2000	25 (projected)	0.36

Source: El-Bushra (1974/75, p.97)

Table 4.3: Sudan - distribution of towns by size class
(1955-1983)

Size Class	Number of towns in:			
	1955	1965	1975	1983
<5,000	33	43	23	25
5,000+	35	64	93	114
10,000+	20	33	50	65
20,000+	10	13	23	35
50,000+	3	7	14	14
100,000+	1	2	6	7

Source: derived from 1955/56 Census; 1964/65
Population & Housing Survey; 1973 Census;
1983 Census

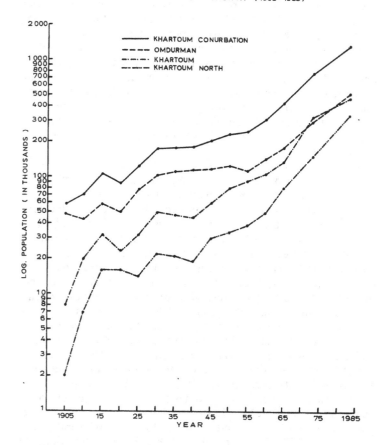

Figure 4.3: Growth of population in the Khartoum Conurbation, 1905-1983

survey conducted in 1980, it was found that 50 per cent of the population of the Khartoum conurbation had been born outside the urban area, and when only those in the age-group above 15 years were considered, the proportion went up to 60 per cent. The survey also showed that over 65 per cent of the immigrants planned to settle permanently in the urban complex (Nour, 1986). Estimates for 1987 show that about 50 per cent of the urban population in Sudan is now concentrated in the triple capital. However, as Sudan's urban system is dominated by a few principal cities, smaller and medium-sized towns recorded much lower rates of growth, generally not exceeding 3 per cent, being similar to the rate of increase prevailing in the rural areas. In fact, some of these smaller towns actually witnessed a decline in population between the 1973 and 1983 Censuses (Table 4.4), as shown in Figure 4.4.

Although smaller and medium-sized towns are considered vital to the process of development in a predominantly agricultural country like Sudan, in many ways they are as underdeveloped as the rural areas in which they are located. Their economic base is weak and less diversified, and their infrastructural networks are far poorer than those found in the major cities. This is perhaps the most conspicuous weakness in Sudan's urban system. In a country where 80 per cent of the population is engaged in agriculture and nomadic pastoralism, the role of small towns is crucial to the socio-economic transformation of the rural areas. These small towns should not only act as collecting and distributing centres, but also as centres for the diffusion of new ideas and technologies throughout the countryside. However, it is doubtful whether these small underdeveloped urban centres are capable of performing such a positive role in the development of these areas. Indeed, several studies of town and country relationships in developing countries point to the parasitic role of cities and their exploitation of rural areas (El-Bushra, 1974/75; Rondinelli, 1983; Salih, 1982; Connell and Curtain, 1982).

The analysis of urban development at the regional or provincial levels reveals a considerable measure of variation. To begin with, natural resources in the form of fertile lands and reliable sources of water, which are considered essential for economic development in Sudan, are unevenly

Table 4.4: Sudan - population change in selected towns, 1973-1983

Town	Total Population 1973	Rank	Total Population 1983	Rank	Percentage Change 1973-1983
Khartoum	333,906	1	476,218	2	43
Omdurman	299,399	2	526,287	1	76
Khartoum North	150,989	3	341,146	3	126
Port Sudan	132,632	4	209,938	4	58
Wad Medani	106,715	5	181,415	5	70
Kassala	99,652	6	142,909	6	43
El-Obeid	90,073	7	139,446	7	55
El-Gedaref	66,465	8	119,002	8	79
Atbara	66,116	9	73,009	13	10
Kosti	65,404	10	91,946	9	41
Nyala	59,583	11	82,972	12	39
Juba	56,737	12	83,787	11	48
Wau	52,750	13	58,008	14	10
El-Fasher	51,932	14	84,533	10	63
El-Geneina	35,424	15	55,996	15	58
Malakal	34,894	16	33,738	20	-3
Sennar	28,546	17	42,803	16	50
Ed Dueim	26,257	18	38,606	17	47
En Nahud	26,005	19	29,787	21	15
New Halfa	24,373	20	37,232	18	53
Shendi	24,161	21	34,505	19	43
El-Gezira Aba	22,218	22	19,632	22	-12

Source: Department of Statistics, Sudan Second Population Census 1973, vol.4, part 2, Khartoum (July 1977); Unpublished Data 1983 Census

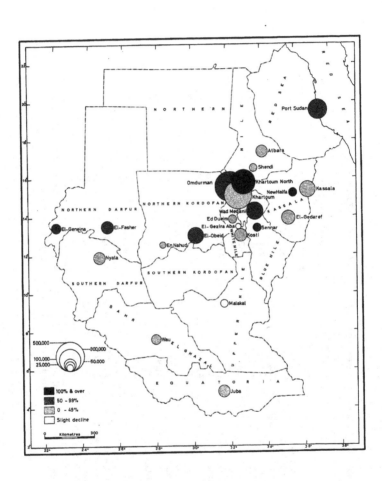

Figure 4.4: Population change in selected towns of Sudan, 1973-1983

distributed. In a country dominated by arid and semi-arid climates, the Nile corridor forms the main focus of economic activity. Moreover, some of the most fertile agricultural lands in Sudan are to be found in the Gezira, the land between the Blue and White Niles (El-Bushra, 1967). The Gezira is also located at the region of intersection between the two most important zones of population scatter, namely, Central Sudan and the Nile Valley, thereby forming the core of the Sudanese ecumene. The urban-dominated Khartoum Province together with parts of Kassala and the Blue and White Nile Provinces, where development projects are located, are part of this central core area (Roden, 1974; Wadehn, 1981; Bronger, 1985) (see Figure 4.4). It is here in the land between the two Niles that the first commercialised agricultural scheme - the Gezira - was introduced in 1925, during the colonial period (see Figure 4.5). Most of the other major agricultural projects are also to be found in this core area, namely the Rahad Agricultural Scheme, the Mechanised Crop Production Schemes of the Gedaref District, the Khashm el-Girba Scheme, the agricultural projects of Damazin in the Southern Blue Nile Province, and the White Nile agricultural schemes (Figure 4.5). It is also worth mentioning that all the dams which were built for the purposes of irrigation and the generation of electricity are located in this core area of Sudan. Concentrated in it are not only the principal agricultural projects, but also the major industrial establishments, the most advanced infrastructural facilities and the largest cities (Figure 4.4). Most of the industrial development in Sudan, whether it is urban or rural-based is located within this core area. The concentration of industries is exceptionally high in the Three Towns conurbation and most of the rural industries such as cotton milling, cotton textiles and sugar crushing are to be found there (El-Bushra, 1972b, 1985). The convergence of the contemporary Sudanese transport network on the core region is well illustrated in Figure 4.6.

As revealed by the 1983 Census, Khartoum Province with 75 per cent of its total population classified as urban, has the highest concentration of urban dwellers in the country, while the Upper Nile Province with 3 per cent was the least urbanised (Table 4.1). The proportion of urban population in the Khartoum region is expected to increase dramatically

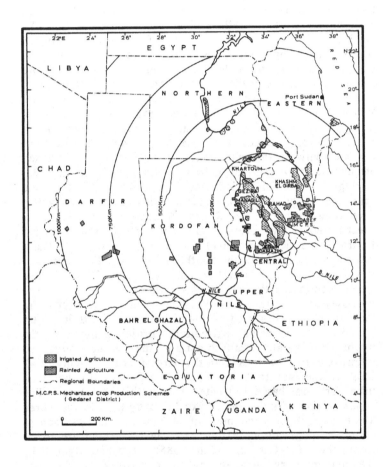

Figure 4.5: Modern agricultural developments in Sudan, 1985

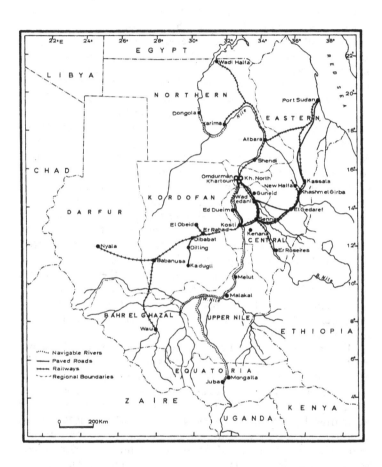

Figure 4.6: The transport network of Sudan, 1987

in the near future as it is anticipated that more rural people will settle in the urban areas. Sudan's triple capital dominates the country demographically, economically, culturally and politically (El-Bushra, 1976b, 1979, 1985/86). Governmental, economic and service functions are highly concentrated in the primate city. In an urban complex with only 6 per cent of the national population (1983 Census), over 75 per cent of the country's industrial establishments, 65 per cent of its banking and commercial companies, 57 per cent of the physicians, 21 per cent of hospital beds, and 90 per cent of enrolments in higher education are to be found (El-Bushra, 1975a, 1975c, 1979; Hijazi, 1981; Nour, 1986). Because of the over-concentration of social and economic functions, the Khartoum conurbation has come to exert a strong pull on the rural masses. Unfortunately, there are no up-to-date figures to indicate the existing pattern of migration to the triple capital. The 1983 Census data have not been released, but it seems likely that the pattern of migration shown by the 1973 Census will have changed substantially from that depicted in Figure 4.1. Specifically, the majority of migrants now come from the southern and western regions, the two most backward areas of the country. Some observers put the number of southern and western migrants to the Three Towns conurbation during the past decade at one million.

The Eastern, Northern and Central regions with 20-30 per cent of their populations being classified as urban in 1983 are second after Khartoum in the degree of urbanisation. The Eastern region has experienced a certain measure of economic development in the form of two important agricultural schemes, namely, the Mechanised Crop Production Schemes and the Khashm el-Girba project mentioned previously. This region contains Sudan's only seaport, and is dominated by three major urban centres. According to the 1983 Census, the population of Port Sudan is in excess of the 200,000 mark, and those of both Kassala and Gedaref 100,000. In addition to internal migration, this region receives a large number of refugees from neighbouring Ethiopia. The Central region which consists of the Blue Nile, White Nile and Gezira Provinces is by far the most advanced agriculturally and has therefore developed a large number of small and medium-sized service centres.

By contrast, the Western region which is composed of the provinces of Kordofan and Darfur and the whole of the Southern region consisting of Equatoria, Bahr el-Ghazal and Upper Nile are the least developed economically and the least urbanised. The economy of the Western region is dominated by traditional agriculture and nomadic pastoralism both of which are totally dependent on adqute rainfall. Fluctuations in the amount of rainfall which have characterised the region for a long time have resulted in crop failures and the death of many animals. Lack of reliable sources of water has retarded economic development and constrained urban growth even though the region accounts for 30 per cent of the country's total population. Although several areas in Sudan have been affected by drought and famine, the Western region has been an area of particular distress. Waves of migrants from the Western region have moved to various rural and urban centres including Khartoum. It was estimated that in 1985, at least 650,000 people had migrated from Northern Darfur alone because of drought and famine (Holcombe, 1987). Major cities in this region include El-Obeid, El-Fasher, Nyala and El-Geneina (Figure 4.4). Similar to the Eastern region, the Darfur Province receives a large number of immigrants from neighbouring Chad.

Because of its physical and cultural isolation Southern Sudan is one of the least developed regions in the world and therefore also one of the least urbanised (El-Bushra, 1967). Torn by armed rebellion and disease, the region remains the most backward in the country economically and culturally, and is presently witnessing a mass exodus of people into the northern areas (Barbour, 1964; Roden, 1974; El-Bushra, 1975b). The Southern region accounts for 25 per cent of the country's population (1983 Census), but has less than 10 per cent of its urban dwellers. The proportion of urban population within the region is less than 8 per cent, implying a similarly low level of economic and infrastructural development. Major Nilotic tribes such as the Dinka, Shulluk and Nuer who inhabit the region are among the most primitive in the world. The main occupations of these tribes are subsistence farming and animal husbandry. Economic life centres around the cattle which are usually driven over considerable distances between the rivers and pastures in their vicintiy (El-Bushra, 1967). As such, it may be more appropriate to classify the Nilotic tribes under the nomadic groups rather than the sedentary population as has been the case in previous censuses, specially those of 1955/56, 1973 and 1983.

Being plagued by the civil war since independence, the Southern region has not had the chance to benefit from the development of production projects. The present armed rebellion has brought to a halt the completion of two major schemes, namely, the Jonglei Canal and the commercial exploitation of petroleum deposits discovered by the Chevron Oil Company. Some observers believe that the discovery of oil might have triggered the present hostilities in the struggle for political power over the region. The present fighting is not only leading to loss of life and property, but will also serve to delay the development of this backward and isolated region for years to come. As a result, the rate of out-migration from the region seems likely to be stepped-up, leading to further regional imbalance at the national level. The Southern region, like the Eastern and Western ones, receives refugees from Uganda because of the internal conflict between the warring factions in that country. The movement across the Sudanese borders of political refugees from neighbouring Ethiopia, Uganda and Chad makes these frontiers active lines of tension for Sudan. Sudan has one of the longest international borders, involving eight countries, and has throughout history been the recipient of foreign migrants.

The previous discussion has demonstrated that Sudan, with 20 per cent of its population classified as urban in 1983 (Table 4.1), is one of the least urbanised countries in the generally poorly urbanised continent of Africa. As an underdeveloped country suffering from declining production and unfavourable terms of trade, Sudan is also unlikely to figure high in urbanisation in the foreseeable future (Table 4.2). However, the existing polarisation of socio-economic activities may be further accentuated, thereby leading to the phenomenal growth of major urban centres, in which case the contradictions between core and periphery will be deepened (Roden, 1974; El-Bushra, 1975b, 1985; Salih, 1982; Babiker and Arbab, 1987) (Tables 4.3 and 4.4). In the absence of a more integrated urban system, the prevailing levels of urban primacy and urban bias are likely to be perpetuated. On the whole, the migratory currents between the developed core, represented by the principal cities and modern agricultural schemes, and the underdeveloped peripheral areas are expected to become more pronounced in the future. Accordingly, in the following section, the causes and consequences of international and internal migrations will be discussed in detail.

International Migration

Perhaps the question of migration is best put into perspective if it is realised that Sudan has always represented a zone of easy movement and consequently, has attracted human migrants from the various parts of Africa throughout time (Hassoun, 1952; Davies, 1964; El-Bushra, 1967; Duffield, 1980). Part of the Muslim tradition in West Africa is to move east across Sudan towards the holy places of Islam in Arabia. As Sudan has offered better opportunities for living than the original homelands of the migrants, large numbers of West Africans have been tempted to settle, so that by the time of the first census in 1955/56, the number of those settling in the country was estimated at over 600,000 out of total population of 10.5 million (Davies, 1964). That figure was considered to be a low estimate for West African settlers whose present numbers are put at more than 3 million. Although most of the settlers have remained in the rural areas, some have come to concentrate in major cities such as Khartoum, Kassala, Port Sudan and El-Obeid, living in shantytowns and providing a pool of cheap labour.

In more recent years, Sudan has been witnessing international migration at a scale never known before. First of all, adverse economic conditions which have prevailed for the past two decades have forced a large number of northern Sudanese to emigrate to the neighbouring oil countries of the Middle East, particularly Saudi Arabia. Although no reliable statistics are available, some estimates put the number who have migrated to the oil-rich countries at more than one million (Weaver, 1985). Other observers believe that the figure could easily reach two million workers. The emigration of professionals, skilled and unskilled workers in such numbers has created a vacuum which has been filled by other groups, and has at the same time adversely affected the country's development schemes. The last decade has also witnessed human migrations on a grand scale into North and Central Sudan, from both internal and external sources. As has been mentioned before, these migrations have been instigated by environmental, economic and political factors. Externally, political instability and civil wars in neighbouring countries such as Ethiopia, Uganda and Chad together with the spread of drought and famine have caused thousands of people to move into Sudan. As estimated by Weaver (1985), as many as 200,000 Ethiopian refugees who were fleeing the civil war came to settle in towns and cities in Northern Sudan, while the number of Ethiopians entering Sudan because of the conditions of drought and famine during the years 1982-1985 was put at about 500,000. According to some observers these figures are on the conservative side and the number of economic and political migrants from Ethiopia alone may be put at 1.5 million (Al-Saiyyad Research and Studies Centre, 1985). Although most of the Ethiopian migrants have come to be concentrated in the Eastern region in the towns of Kassala, Gedaref and Port Sudan, large numbers have settled in the Khartoum urban area. In the same way, some of the migrants from Uganda (250,000) and Chad (123,000) have settled either in the regions adjacent to their countries or else have headed towards the triple capital (Ministry of Interior, 1985). There is no doubt that the movement of more than two million people during the past few years from neighbouring countries into Sudan has caused a lot of strain on an economy already suffering from many difficulties. Sudan's debts, as estimated by the World Bank, are put at more than US $ 10 billion, and the balance of payment deficit at about one billion US dollars (Davies, 1986). In a crippled economy like that of Sudan, the massive movement of foreign migrants into the country is looked upon as an extra heavy burden to the national economy. The influx of large numbers of refugees has been at least partially responsible for inflation (now standing at 35 per cent), periodic shortages of goods and overcrowding of social services. In the urban centres, thousands of foreign migrants compete with Sudanese nationals for jobs and facilities such as schooling, housing and medical care. Increasing demand on a rather limited supply of goods and services has led to exorbitant food prices and rents, with many dealings taking place on the black market.

Internal Migration

Similar to international migration, internal migration has during the past decade been stimulated by economic, political and environmental factors. Sectoral development or growth pole strategies, which have characterised economic life since the creation of the modern state at the turn of this century, have served to increase the disparities between core and periphery, as noted earlier. The movement of people from the underprivileged areas to the core has been active during the past two decades and there is no sign of this slackening in the near future. The concentration of people into the riverine core leads to more investment programmes, thereby leading to greater regional inequalities and discontent in the economic backwaters (Roden, 1974; El-Bushra, 1985; Davies, 1986). The dual economy of Sudan is composed of a small modern sector in which most of the production is geared towards foreign markets, and a large traditional sector in which most of the production is channelled into local markets. Recent population movements together with managerial and financial difficulties have resulted in a decline in production of both food and cash crops. The emigration of large numbers of young men from the traditional areas, where over 60 per cent of Sudan's population is located, has adversely affected the level of production, particularly that of food staples, so that Sudan is now a net importer of foodstuffs. Equally, the modern sectors have witnessed sharp declines in production levels, because of managerial and financial problems (Davies, 1986). This depressing economic situation has deteriorated still further with the spread of droughts in the period 1983-1985, and the intensification of the armed rebellion in southern Sudan. The war in the Southern region is costing the Central Government about US $ 100 million a year, a sum of money badly needed for development purposes (Al-Saiyyad Research and Studies Centre, 1986).

In reviewing the causes and consequences of rural-urban migration at the regional and provincial level, some or all of the above mentioned factors may be operational, and the degree of variation between one region and another is a reflection of the level of development attained in each region. The Khartoum region with a density of 86 persons per square kilometre in 1983 is the most densely settled in the country and is completely dominated by the Three Towns conurbation which accommodates 75 per cent of the population of the region. The remaining 25 per cent are engaged in market gardening and animal production to supply the urban markets. Rural-urban migration is still active in the region which is by far the most developed in the country. The primacy of the urban complex has already been underlined, and the real or imagined opportunities for employment keep the rate of migration into the triple capital at a high level. The economic base of the Khartoum metropolitan area is rather weak and underdeveloped, reflecting the country's adverse economic conditions. Basic infrastructural facilities such as water, electricity, sewerage, roads, health-care and garbage collection services have been completely paralysed in more recent years because of population pressure, planning and managerial problems and financial difficulties. Industrial establishments which are highly concentrated in this urban agglomeration have been working recently at no more than 25 per cent of their full capacity because of power cuts and the lack of availability of spare parts due to an acute shortage in foreign exchange (Davies, 1986; El Sammani *et al.*, 1986).

Mass migration of rural people into the capital city has been associated with the creation of shanty towns of unprecedented magnitude. During the past decade more than one million people have congregated into the urban complex with the result that huge squatter settlements have now come to litter the peripheral areas of these large cities as shown in Figure 4.2 (Krings, 1978). At least 40 per cent of the population of the triple capital lives in substandard dwellings or in shanty towns where there is an almost complete lack of basic amenities. These form a constant health hazard to the city, and diseases such as malaria, leishmaniasis and typhoid, which were once confined to isolated pockets in Southern and Western Sudan, are now endemic in the Khartoum urban

area (El-Bushra, 1976c, 1979/80; El-Sammani *et al*., 1986).
Migrants from Western and Southern Sudan form the
majority of squatters in these shanty areas and present the
local authorities with a range of problems. Almost all
squatter settlements which tightly ring the urban areas are
built on privately-owned land. In an underdeveloped urban
economy like that of the triple capital, the rate of
unemployment among the workforce is high, and of course,
this is particularly so among the migrant population.

Although there are no up-to-date figures, it was found as
early as 1974, that 20 per cent of all men in the age-group
15-19 years in the Khartoum conurbation were unemployed,
together with 11 per cent of all men in the 20-24 age group
(Galal-Al-Din, 1980). At present masses of unemployed
migrants have either opted for trivial trades or have turned
to begging. Hundreds of venders selling small merchandise
such as ready-made clothing, cigarettes, vegetables and fruit
together with car washers and shoeblacks crowd the central
business districts of the Three Towns. The past decade has
also witnessed, for the first time, the emergence of a new
group of homeless people who do not even own a shack in
the squatter areas, and who have been forced to take to the
pavements as their living quarters.

One of the most outstanding negative outcomes of this
large-scale migration of people into the triple capital has
been a serious deterioration in the level of security. Thus,
although the overall rate of crime in the urban complex went
up by 8 per cent during the period 1984/85, homicide cases
increased 40 by per cent. Moreover, during 1985, the capital
city's share of the national total was 20 per cent in the case
of armed robberies, 22 per cent for breaking and entering,
40 per cent of thefts, 46 per cent of the 3,693 cases of drug
dealing, and 55 per cent of total traffic accidents (Ministry
of Interior, 1985). Ironically, polarisation of socio-economic
activities in the urban area has given birth to a similar over-
concentration of social ills. In order to alleviate the existing
difficulties and pressures, the authorities need to take steps
to curb the influx of rural people into the capital city.
However, as the authorities lack the ways and means, the
imagination, and the initiative to tackle the existing problems
at both the rural and urban levels, the future of the national
capital seems bleak.

The Central region which is made up of the provinces of Blue Nile, White Nile and Gezira is the second most developed in the country after Khartoum and the greater part of this region together with Khartoum conurbation constitutes the riverine core. In addition to the important agricultural schemes of Gezira, Managil and Raha, which form the backbone of the Sudanese economy, major industrial establishments such as cotton ginning and textiles and sugar crushing are located in this region. As such, the Central region receives a lot of migrants from the marginal areas, particularly from the Western region. The seasonal labour required for the various agricultural operations has always been provided by western migrants. The principal cities of this region are Wad Medani, Kosti and Ed Dueim of which Wad Medani is the largest and fastest growing (see Figure 4.4). Dense agricultural settlements in this region necessitated the development of a large number of small, although ill-equipped service centres.

High rates of natural increase, widespread unemployment and underemployment and the spread of primary education, which is not oriented to serve rural development, have all been important factors contributing to the emigration of young people from the rural areas. As the region is the nearest and best connected with the capital city, the bulk of migrants have headed towards the urban complex (El-Bushra, 1976b). For many migrants, settlement in the Khartoum urban area, even under the most adverse conditions, gives hope of finding a job, if not in the triple capital then in one of the oil-rich countries of the Arabian Gulf. So while western migrants focus on this region either for seasonal or permanent agricultural work, the original inhabitants seek employment elsewhere. The emigration of large numbers of young people from this region means that agricultural work has to depend on hired labour, and therefore peasants have to settle for a smaller profit margin. According to the 1983 Census, the Central region has the third highest share of the country's population (20 per cent), and is also third with respect to its level of urbanisation (20 per cent) (Table 4.1). However, except for a limited movement of rural people to the major cities of Wad Medani, Kosti and Ed Dueim, the majority of migrants move to the triple capital. This is a clear indication that the socio-economic and infrastructural facilities found in the towns of this region are by far inferior to those of the Khartoum conurbation and as such, the regional centres become less attractive to migrants. So in spite of the fact that the region has the second highest population density in the country after Khartoum with 30 per square kilometre, the degree of interaction between urban and rural areas is relatively low.

The Eastern region with 29 per cent of its population living in urban centres, is the second most urbanised after Khartoum, but has the highest concentration of nomads, standing at some 25 per cent (Table 4.1). The major production projects of the Gedaref Mechanised Crop Production Scheme, the Khashm el-Girba Scheme and various industrial projects at Port Sudan and New Halfa are considered to be part of the developed core. Urban life in this region is dominated by three principal cities, namely, Port Sudan (ranking second in the country's urban hierarchy after Khartoum), Kassala and Gedaref. These three urban centres are all located in areas of agricultural and industrial production and are growing quickly. They each receive migrants from the surrounding region, as well as from far-distant ones, and even from abroad. From within the region, nomadic groups such as the Beja focus on Port Sudan for work in the docks, and the Shukriya concentrate in Gedaref town. The availability of agricultural and industrial work in this region has attracted some migrants from the Southern region and large numbers from the Western region and West Africa. People of West African origin are to be found in all areas of agricultural production, as well as having a strong representation in all major cities (Duffield, 1980).

As mentioned earlier, in recent years all of the big towns of this region, but particularly Kassala, have been overwhelmed by Ethiopian refugees. Kassala, being a frontier town, has always received Eritrean migrants from Ethiopia. The volume of foreign migration to this town is such that it is leading to acute shortages in the supply of goods and services, and has brought havoc to the once neat and prosperous town. The urban environment has deteriorated rapidly, with the outbreak of epidemics constantly possible, while the rate of crime has taken a sharp upward turn, especially armed robberies and homicide (Ministry of Interior, 1985). Existing sectoral development in the region is expected to continue, resulting in even greater urban bias and a further stream of rural-urban migration.

The Northern region which is made up of the Nile and Northern provinces has a long tradition of out-migration. Being endowed with meagre natural resources, high rates of population increase and a well-established educational system (Galal Al-Din, 1980), the Northern region has never been particularly attractive to its inhabitants. The prevailing desert conditions have made life cling very strongly to the Nile, and the shortage of agricultural land has ruled out the chance of large-scale agricultural development. For quite a considerable time the population of this region has shown high rates of out-migration, not only to the other areas of Sudan, but also to neighbouring countries such as Egypt and Saudi Arabia. The majority of the population of the region live in rural sedentary habitations along the Nile (74 per cent), with 21 per cent living in small urban centres. The largest towns are Atbara, the headquarters of Sudan Railways, and Shendi, both of which are located in the Nile Province (see Table 4.4). Currently, most of the migrants from this region move internally towards Khartoum and Port Sudan, and externally to Saudi Arabia.

Heavy emigration of young people of employable age from this region has been disruptive economically, socially and demographically. In many villages, those left behind are the females, children and old people, who contribute little directly to economic production. Many people, even whole villages in this region, have become highly dependent on remittances sent to them by family members working elsewhere. Remittances have in most cases little impact on local production, as the money received is normally used for consumption, going toward housing, clothing and foodstuffs, rather than being invested in agriculture. In fact, the very existence of a guaranteed source of income has resulted in some agricultural land being removed from production. In rare instances, remittances have been used individually or collectively to modernise and develop agriculture and to provide water, electricity and transport in some villages and small towns (Van Western and Klute, 1986). In this economically underdeveloped region, the degree of interaction between urban and rural areas is rather limited and both urban and rural population engage in out-migration.

The Western region, one of the least developed in the country, has about 10 per cent of its population living in urban centres, but over 20 per cent are classified as nomads (Table 4.1). Major towns in this region are El-Obeid, El-Fasher, Nyala and El-Geneina, all of which receive some migrants from the rural areas, with both El-Fasher and Nyala located in the western-most province of Darfur receiving migrants from neighbouring Chad. In this region where 30 per cent of the population is concentrated, there are hardly any modern development projects, and the economy is dominated by traditional farming and animal husbandry. Widespread subsistence economies together with the recent drought and famine have, as previously stated, made this region one of the important areas of out-migration.

Migrants from the Western region are internally strongly represented in all the production areas in the river core, and externally in the Arabian Gulf countries and Libya. Western migrants usually head for modern areas of production within Sudan and after learning new skills, either move to Khartoum or abroad. Generally, it has been found that the learning of new skills and the attainment of even a primary level of education increase significantly peoples' propensity to migrate. The large-scale emigration of active people from this region indicates a further deterioration of the economic situation.

The Southern region, which forms the real backwater of the Sudenese economy, faces considerable problems, as previously mentioned. Adverse economic and political conditions have forced thousands of southerners to move northward into the riverine areas. With little experience in agriculture, southern migrants have come to concentrate in major cities such as Khartoum, Kosti, Wad Medani and New Halfa. It has been estimated that during the past decade, at least 500,000 southerners must have moved into the Khartoum urban area alone. Although the emigration of large numbers from the Southern region is bound to lead to further deterioration of the economic situation, two positive outcomes have resulted. One is that for the first time, the people of the North and South have been brought into direct contact with one another, a step forward towards national reconciliation and cohesion.   The second result has been that by being in the North, southerners have come to learn new skills as well as adopting Arabic as their main language of communication with the northern Sudanese.   However, the emigration of southerners to the neighbouring Arab countries has been relatively slight during the past, due mainly to the language barrier.   Within the Southern region itself interaction between urban and rural areas is very limited, except for the rural-urban migration that has been forced by the civil war.   If the fighting continues for a long period and more rural areas are exposed to it, an exodus to the local towns, which are being protected by government forces, may lead to a sudden upsurge in urban population (El-Bushra, 1975c).   Southern towns are generally service centres, but lack commercial and industrial development, and as such, will continue to be less attractive to migrants than those located in the northern part of the country.

Conclusion

The pattern of Sudanese urbanisation is similar to that of other underdeveloped countries, being characterised by urban primacy and a poorly integrated urban system.   Urban primacy at the regional level is far less pronounced than that of the national capital.   Sudan's urban networks are entirely dominated by the Three Towns conurbation, which presently accommodates about half the nation's city-dwellers.   The domination of urban life by a few principal cities has minimised the role of smaller towns in the social and economic transformation of the rural areas.   As the economic and infrastructural base of small towns is weak and underdeveloped, these centres are becoming less and less attractive to the rural people.   Sectoral development or growth pole strategies, which have constituted the adopted development policy during the colonial and post-independence periods, have manifestly failed to bring about the required socio-economic progress.   Indeed, the policies adopted have come to underline the contradictions between core and periphery, and have emphasised regional inequalities to the point of discontent and even rebellion in the underprivileged areas. The net result of all this has been a massive movement of people from the less-favoured areas to the more developed centres, particularly the primate city. Heavy migration from rural to urban centres has been disruptive at both ends, leading to a deterioration in productive capacity and service provision. Urban and rural deprivation and poverty, and deficiencies in basic facilities have become widespread. Piecemeal planning, which has characterised various governments in their efforts to solve rural and urban problems, is by no means a way out of the present dilemma. Urban and rural problems cannot be tackled separately, rather a single comprehensive approach is bound to be more effective. Thus, the long term solutions to the existing problems are vested in comprehensive planning, where modernisation and development are considered simultaneously both from below and above. As more development programmes are introduced in rural and small urban centres, the disparities between the regions will be reduced. The authorities should concentrate more on the proliferation of small scale development projects in an attempt to modernise the urban and rural sectors. The economic and infrastructural bases of small and medium-sized towns should be strengthened, so that these towns can come to play their full role in the process of development. Equally, the standard of facilities such as transport, water and electricity should be improved substantially in large regional centres in order to assist the development of a strong industrial base. In this way, local and regional centres will be able to absorb most of the rural migrants, thereby reducing the pressure on the national capital. In the case of the Khartoum conurbation, the development and modernisation of cities such as Wad Medani, Ed Dueim and Shendi which are located more than 150 kilometres away, helps to absorb at least part of this uncontrolled migration to the triple capital.

Moreover, one important area where the local authorities can play a leading role in the process of development is by encouraging and organising people's self-help efforts. Self-help programmes which aim to improve economic production and the provision of community services have been successful in both rural and urban areas. People have come to realise that by pooling their resources together they are able to solve some of their problems. In this way, the government and people can cooperate in an effort to eradicate problems of deprivation and poverty, and at the same time, share in the products of modernisation and development.

REFERENCES

Al-Saiyyad Research and Studies Centre (1985) 'The new Sudanese government and what it inherited from Numeri's regime', *Reports and Backgrounds* (Arabic), Beirut, 1-12

Al-Saiyyad Research and Studies Centre (1986) 'Southern Sudan - negotiated or military solution', *Reports and Backgrounds* (Arabic), Beirut, 1-11

Babiker, A.A. and Arbab, M.I. (1987) 'The problem of regional imbalance in the Sudan and its impact on national cohesion', unpublished paper presented at the *Second International Symposium on the Nile Basin*, 1-7, March 1987, (Arabic), Cairo

Barbour, K.M. (1964) 'North and South in the Sudan, a study in human contrasts', *Annals of the Association of American Geographers* 54, 209-26

Bronger, D. (1985) 'Metropolitisation as a development problem of Third World countries: a contribution towards a definition of the concept', *Applied Geography and Development* 26, 71-97

Connell, J. and Curtain, R. (1982) 'The political economy of urbanisation in Melanesia', *Singapore Journal of Tropical Geography* 2, 119-36

Davies, H.R.J. (1964) 'The West African in the economic geography of Sudan', *Geography* 49, 222-35

Davies, H.R.J. (1986) 'The human factor in development: some lessons from rural Sudan', *Applied Geography* 6, 107-21

Duffield, M. (1980) 'West African settlement and development in the towns of Northern Sudan', in: Pons, V. (ed.), *Urbanisation and Urban Life in the Sudan*, Department of Sociology and Social Anthropology, University of Hull, 209-46

El-Agraa, O.M.A. *et al.* (1985) *Popular Settlements in Greater Khartoum*, Silver Star Printing Press, Khartoum

El-Arifi, S.A. (1980) 'The nature and rate of urbanisation in Sudan', in: Pons, V. (ed.), *Urbanisation and Urban Life in the Sudan*, Department of Sociology and Social Anthropology, University of Hull, 381-411

El-Bushra, E.S. (1967) 'The factors affecting settlement distribution in the Sudan', *Geografiska Annaler* 49, B, 10-24

El-Bushra, E.S. (1969) 'Occupational classification of Sudanese towns', *Sudan Notes & Records* 50, 75-96

El-Bushra, E.S. (1971a) 'The evolution of the Three Towns', *African Urban Notes* 6, 8-23

El-Bushra, E.S. (1971b) 'Towns in the Sudan in the eighteenth and early nineteenth centuries', *Sudan Notes and Records* 52, 63-70

El-Bushra, E.S. (ed) (1972a) *Urbanisation in the Sudan*, Proceedings of the 17th Annual Conference of the Philosophical Society of the Sudan, 2-4 August, Khartoum

El-Bushra, E.S. (1972b) 'The development of industry in Greater Khartoum, Sudan', *The East African Geographical Review* 10, 27-50

El-Bushra, E.S. (1974/75) 'Urbanisation in the Sudan', *Bulletin de la Société de Geographie d'Egypt* 47/48, 95-103

El-Bushra, E.S. (1975a) 'Sudan's Triple Capital: morphology and functions', *Ekistics* 39, 246-50

El-Bushra, E.S. (1975b) 'Regional inequalities in the Sudan', *Focus* 26, 1-8

El-Bushra, E.S. (1975c) 'The Sudan', in: Jones, R. (ed.), *Essays on World Urbanisation*, George Phillip, London, 377-389

El-Bushra, E.S. (1976a) *An Atlas of Khartoum Conurbation*, Khartoum University Press, Khartoum

El-Bushra, E.S. (1976b) 'The sphere of influence of Khartoum conurbation, Sudan', *Erdkunde* 30, 286-94

El-Bushra, E.S. (1976c) 'The distribution of population and hospital facilities in the Sudan', *Sudan Medical Journal* 14, 78-81

El-Bushra, E.S. (1979) 'Some demographic indicators for Khartoum conurbation, Sudan', *Middle Eastern Studies* 15, 295-309

El-Bushra, E.S. (1979-80) 'Some environmental problems of desert towns: the case of Khartoum and Riyadh', *Bulletin of Arab Research and Studies* 10, 83-95

Urban Crisis in Sudan

El-Bushra, E.S. (1985) 'Development planning in the Sudan', *Erdkunde* 39, 55-59
El-Bushra, E.S. (1985-86) 'Greater Khartoum: evolution and development, *Bulletin of Arab Research and Studies* (Arabic), 12, 5-28
El-Sammani, M.O. *et al.* (1986) *Management Problems of Greater Khartoum*, Institute of Environmental Studies, University of Khartoum
Farghali, A.M.A. (1977) *Rapid Urbanisation and Low-Income Housing in the Sudan: a study of the Three Towns*, unpublished M.Phil thesis, University of Edinburgh
Galal-Al-Din, M.A. (1980) 'The nature and causes of labour migration to Khartoum conurbation, in: Pons, V. (ed.), *Urbanisation and Urban Life in the Sudan*, Department of Sociology and Social Anthropology, University of Hull, 425-448
Gilbert, A. and Kleinpenning, J. (1986) 'Migration, regional inequality and development in the Third World', *Tijdschrift voor Economische en Sociale Geografie* 77, 2-6
Hale, G.A. (ed.) (1971) *Sudan Urban Studies*, a special issue of *African Urban Notes*, 6, 2, Michigan State University
Hamdan, G. (1960) 'The growth and functional structure of Khartoum', *Geographical Review* 50, 21-40
Hassoun, I.A. (1952) 'Western migration and settlement in the Gezira', *Sudan Notes and Records* 33, 60-112
Henin, R.A. (1962) 'Economic development and internal migration in the Sudan', *Sudan Notes and Records* 44, 100-19
Herbert, D.T. and Hijazi, N.B. (1984) 'Urban deprivation in the developing world - the case of Khartoum/Omdurman', *Third World Planning Review* 6, 263-81
Hijazi, N.B. (1981) *Urban Structures and Socio-Spatial Problems in Khartoum Urban Area: a geographical study*, unpublished Ph.D thesis, University of Wales
Holcombe, A. (1987) 'Reaching beyond famine relief: planning strategy for rehabilitation and development in Darfur region, Western Sudan', *Geojournal* 14, 11-18
Krings, T. (1987) 'Surviving in the periphery of the town - the living conditions of Sahelian drought refugees in Mopti (Republic of Mali)', *Geojournal* 14, 63-70
Lewis, J.R. (1986) 'International labour migration and uneven regional development in labour exporting countries', *Tijdschrift voor Economische en Sociale Geografie* 77, 27-41
Lobban, R.A. (1975) 'Alienation, urbanisation and social networks in the Sudan', *Journal of Modern African Studies* 13, 491-500
Ministry of Health (1984) *Annual Report 1984* (Arabic), Khartoum
Ministry of Interior (1985) *Annual Criminal Report 1985* (Arabic), Government Press, Khartoum
Nour, O.E.M. (1986) 'Rapid urban growth in Greater Khartoum - origin, components and causes', *Arab Journal of Humanities* (Arabic) 6, 84-101
Pons, V. (ed.) (1980) *Urbanisation and Urban Life in the Sudan*, Department of Sociology and Social Anthropology, University of Hull
Roden, D. (1974) 'Regional inequality and rebellion in the Sudan', *Geographical Review* 64, 498-516
Rondinelli, D.A. (1983) 'Towns and small cities in developing countries', *Geographical Review* 73, 379-95
Salih, K. (1982) 'Urban dilemmas in southeast Asia', *Singapore Journal of Tropical Geography* 3, 147-61
Van Western, A.C.M. and Klute, M.C. (1986) 'From Bamako with love: a case study of migrants and their remittances', *Tijdschrift voor Economische en Sociale Geografie* 77, 42-9
Wadehn, M. (1981) 'Urban and regional development in Brazil - the policy of intermediate centres', *Applied Geography and Development* 18, 63-79
Weaver, J.L. (1985) 'Sojourners along the Nile: Ethiopian refugees in Khartoum', *Journal of Modern African Studies* 23, 147-56
Winters, C. (1977) 'Traditional urbanism in the north central Sudan', *Annals of the Association of American Geographers* 67, 500-20

5

RURAL-URBAN INTERACTION AND DEVELOPMENT IN SOUTHERN AFRICA: THE IMPLICATIONS OF REDUCED LABOUR MIGRATION

DAVID SIMON

High levels of urban-rural linkage are found in most sub-Saharan African countries. Indeed, a feature of this continent is the apparently voluntary maintenance of such ties long after decolonisation (O'Connor, 1983). However, southern Africa can be distinguished from other regions of the continent by the extent to which labour migration was institutionalised during the colonial era. The regional political economy was predicated on subjugation and dispossession of the indigenous peoples, who were then forcibly concentrated in small, inadequate reserves or bantustans away from the best agricultural land and usually towards the geographical periphery of the respective territories. Imposition of hut and poll taxes, payable in cash, coupled with the inadequacy of peasant subsistence in the reserves, compelled Africans to seek progressively more wage labour in the colonial money economy.

A highly centralised and repressive labour regime, known as the contract labour system, was engineered to ensure, in the words of the notorious 1917 Stallard Commission, that Africans could remain in towns, mines and on commercial (ie. white-owned) farms only insofar as they 'minister to the white man's needs' (see Rich, 1978). The system was most highly regimented in Zimbabwe (formerly Rhodesia), South Africa, and Namibia (formerly South West Africa) which is still occupied by South Africa (Cronje and Cronje, 1979; Gordon, 1977; Moorsom, 1977a, 1977b; Palmer, 1977; Phimister, 1977; Wilson, 1972). For the most part, women, children and the elderly were forced to remain in the reserves (see Figure 5.1), while urban workers had to return to their rural homes annually as a condition of renewal of

Figure 5.1: The political geography of southern Africa

their contracts. When in town, migrants were accommodated in ultrabasic, crowded and prisonlike hostels, minimising the costs to their employers (Gordon, 1977; Van Onselen, 1976; Wilson, 1972). This was broadly the pattern throughout the region, which Samir Amin (1972) appropriately labelled 'Africa of the labour reserves', although details and the extent of systematisation differed between individual territories and over time. These differences are significant and should not be obscured by simplistic generalisation (Crush, 1984).

The contract labour regime represented a peculiarly institutionalised and inequitable system of urban-rural interaction, based upon urban development and rural underdevelopment. In many ways, South Africa has long held the key to understanding the relationships between labour migration and the political economy of exploitation throughout the region. The first point is historical. As the oldest European colony, with the largest settler population and most sophisticated modern economy, British labour policy there served as a model for much of southern and East Africa. This was true not only of other British dependencies, but also the German colonies of South West Africa and Tanganyika (Bley, 1971). Secondly, expansion and development of the South African economy, particularly its mineral-based heartland on the Witwatersrand, has long relied upon large-scale employment of foreign contract labour drawn from as far away as Malawi and Zambia. Geographically unequal capitalist development has delineated a clear core and periphery within the regional spatial system, which can be defined in terms of economic and political power, infrastructural linkages and, most importantly, also labour migration. Recruitment is highly centralised through The Employment Bureau for Africa (TEBA), formerly known as The Witwatersrand Native Labour Association (WENELA), which is run by the South African Chamber of Mines. For several of the peripheral countries of the region, perhaps most dramatically Lesotho, southern Mozambique and, until the mid-1970s, Malawi, labour export has been, and remains, a vital source of employment and foreign exchange income, thus perpetuating the system long after these countries attained independence and dismantled their own institutionalised labour regimes (Christiansen and Kydd, 1983; First, 1983; Murray, 1981).

Even today, nearly twice as many Basotho men find waged employment in South Africa as in their own country, given the paucity of opportunities outside the peasant sector in that mountainous enclave. The 1982 figures were 140,700 and 85,000 respectively (Republic of South Africa, 1985). Migrants' remittances from South Africa in 1983-4 represented an astonishing 51.9 per cent of Lesotho's GNP, having increased from 33.9 per cent a decade earlier (Wellings, 1986, p.219). Since migrant numbers remained roughly constant over the intervening period (Table 5.1), this major increase reflects rising real mine wages as well as Lesotho's domestic economic crisis. These harsh geographical and economic realities of dependence go a long way towards explaining Lesotho's by and large accommodating political stance *vis-à-vis* its dominant neighbour, despite its membership of the nine-member Southern African Development Co-ordinating Conference (SADCC), which was established in 1979 expressly to reduce dependence on South Africa. It is thus no coincidence that Lesotho, virtually alone, has been little affected by a major policy shift in South African recruitment policy. Prompted by a combination of external and internal circumstances, this resulted in the number of foreign migrants more than halving from 1973 to 1980, in favour of domestic labour (Christiansen and Kydd, 1983; Cobbe, 1986; Crush, 1986; First, 1983; Lemon, 1982; Murray, 1981; Taylor, 1982, 1986).

It would be an oversimplification to suggest that incorporation into the regional political economy totally destroyed indigenous precapitalist social formations or their modes and relations of production. Rather, they were transformed and mutated through what may usefully be conceived of as a process of 'conservation and dissolution', whereby those aspects exploitable by the settler economy were allowed and indeed often encouraged, or even forced, to persist in a suitably dependent manner. An example from the political sphere would be colonial bastardisation of traditional chieftainships into client positions of authority, administration and social control. In economic terms, two related examples are particularly important for an understanding of the system. Firstly, agricultural and

Table 5.1: Numbers of registered foreign African migrant workers in South Africa, 1973–1983

Country of Origin	1973 (June)	1977 (Feb.)	1979 (June)	1980*	1982*	1983*
Angola	42	805	275	291	120	100
Botswana	46,192	43,159	32,463	23,200	26,300	26,000
Lesotho	148,856	160,634	152,032	140,746	140,700	145,800
Malawi	139,714	12,761	35,803	32,319	27,600	29,600
Mozambique	127,198	111,257	61,550	54,424	59,300	61,200
Swaziland	10,032	20,750	13,006	10,377	13,600	16,800
Zambia	684	766	809	864	800	800
Zimbabwe	3,250	32,716	21,547	19,853	11,300	7,700
Others	9,132	9,000	9,224	3,102	2,500	2,500
Total +	485,100	391,848	326,709	285,176	282,300	290,500

Source: Lemon (1982); Republic of South Africa (1985, 1986)

+ includes other nationalities
* month unspecified

infrastructural policy discriminated in favour of the settlers, undermining that section of the African peasantry which had initially flourished in response to the expanding settler economy. Secondly, most Africans were legally compelled to remain locked into sub-subsistence cultivation in the reserves, while becoming increasingly reliant on income from predominantly urban wage labour. Day to day *maintenance* and longer term *renewal* or *reproduction* of urban labour were thus separated both geographically and in terms of the modes of production within which they occurred. This system enabled the payment of low money wages (justified on the grounds that workers had a rural subsistence base), and was also used to perpetuate several politically legitimising myths for the reserve or bantustan system. These held that the traditional, and implicitly viable, peasant mode of production existed in the reserves, that this is what Africans were best suited to, and that the reserves were therefore where they should exercise political rights rather than in the towns and cities. The ultimate objective of this system has been to externalise the cost of reproducing cheap labour power away from urban mining and industrial capital and the state and onto workers and their extended families in the reserves (Beinart, 1982; Bell, 1986; Bundy, 1979; Crush, 1980a, 1980b; Ettinger, 1972; First, 1983; Legassick, 1975; Leys, 1975; Lye and Murray, 1980; Moorsom, 1977a, 1977b; Murray, 1981; Palmer, 1977; Phimister, 1977; Swindell, 1979; Van Onselen, 1976; Wilson, 1972; Wolpe, 1972).

In many parts of rural southern Africa, the situation has become critical as the land base has been progressively undermined through overcrowding, land degradation, drought and famine, the impact of inappropriate state policies on peasant production, and increased rural-urban disparities in income earning opportunities, service quality and availability. While these phenomena are not totally and uniquely attributable to the reserve policy, in that similar problems exist throughout the Third World (Riddell, 1981), they have certainly been greatly exacerbated by it. Furthermore, South Africa's reduced recruitment of foreign migrants since 1973 has precipitated major employment and economic crises for several of its neighbours (Bush *et al.*, 1986; Wellings, 1986).

The objective of this chapter is to review, in this context, two current issues concerning urban-rural interaction: the impact of returned international migrants from South Africa and Zimbabwe, and the national implications of rural development policies involving agrarian reform together with attempts to eliminate migrant labour. These illustrate well the centrality of rural-urban population mobility (most notably institutionalised labour migration) and attendant commodity and financial flows to national and international political economic analysis and development strategies throughout the region.

International migration in southern Africa represents a pervasive form of rural-urban interaction in two important respects. Firstly, as explained above, it has underpinned the historical development of the region's political economy. Any major change in migration could therefore be expected to have repercussions for both the sending and receiving countries, as well as politico-economic relationships between them. Secondly, the vast majority of international migrants come from the rural peasant sector and go to work in mines which are predominantly located in urban areas. Their remittances are often crucial to rural household survival and to national exchequers. On completion of their contracts, migrants must return to the peasant sector, commercial agriculture or the urban areas of their own country. Whether they find alternative wage employment, remain unemployed or are forced to revert to peasant underemployment, is clearly crucial to the respective national economies and to their prospects for political stability and development. Malawi and Mozambique are used as case studies because they have experienced the largest and most rapid reductions in labour exports.

The other focus of this chapter is agrarian reform. In view of the severe land pressure in many rural reserves, agrarian reform and resettlement are prerequisites for equitable development. State policy may also seek to reduce or eliminate migrant labour for moral and equity reasons. However, as the Zimbabwean case study reveals, the measures invoked to achieve these objectives may conflict with policy in other spheres, thus undermining the prospects for success.

The impact of returning migrants on, and implications for, peripheral economies

The Chamber of Mines had long pursued a policy of expanding its labour catchment area in order to keep wages depressed by avoiding direct competition for domestic workers with South Africa's expanding secondary sector, where wages were typically two to three times higher. Recruitment from Lesotho and Botswana more than doubled during the 1950s and 1960s, and that from Mozambique remained high, at around 100,000. Notwithstanding the expulsion of the WENELA from Tanzania and Zambia after their independence, it was from the region north of $22^{\circ}S$, most conspicuously Malawi, that increased recruitment was most marked. This peaked in 1973, at 127,970 or 30.3 per cent of total Chamber of Mines employment, and 139,714 or 28.8 per cent of all foreign African employment in South Africa (Crush, 1986; Lemon, 1982; Taylor, 1982).

The pattern of mine recruitment changed dramatically after 1973, as a result of several coincident factors. On the one hand, rising world gold prices in the early 1970s were in part passed on as wage increases. Real wages thus rose significantly for the first time in some sixty years, making the dangerous underground employment more attractive to South Africans. On the other hand, Malawi suspended South African recruitment of its nationals following an air crash in April 1974 in which seventy-four miners were killed. At this point, many Malawians actually terminated their contracts in order to return home. Despite moderate short term increases in partial compensatory recruitment from the BLS countries (Botswana, Lesotho and Swaziland), along with Mozambique and Zimbabwe, over the next two years, the number of foreign migrants declined rapidly, from over 485,000 in 1973 to just over 285,000 in 1980. Thereafter, numbers appear to have stabilised at roughly that level (Table 5.1; see also Crush, 1986). As a proportion of Chamber of Mines labour, foreign migrants fell from 79.5 per cent in 1973 to 45 per cent in 1978 (Taylor, 1982).

The end of Portuguese colonial rule in Mozambique in 1975 and assumption of power by an avowedly Marxist government hostile to South African policies, was another important factor contributing to these changes. From

Table 5.1, it is evident that apart from Lesotho, which maintained a high level of labour exports, the burden of this decline fell primarily on Malawi and Mozambique, and it is to a consideration of the impact of this that the chapter now turns.

Malawi and Mozambique have several common features: peripheral status within the southern African political economic system, their roles as major labour exporters to South Africa, and serious geographical polarisation of economic development around their capitals and major commercial cities, with underdeveloped northern peripheries. In many other respects, however, these two neighbouring countries could hardly be more different. Malawi, a landlocked but densely populated sliver of land wedged between its three neighbours (see Figure 5.1), has been ruled with an iron fist since independence by the quixotic, conservative octogenarian, Dr. Hastings Kamuzu Banda. He has pursued a pragmatic modernisation strategy, arousing the ire of other SADCC members on account of his unashamed good relations with South Africa. At least superficially, Malawi remains one of the most stable countries in the subcontinent and has attained a high degree of self-sufficiency in food, together with significant economic expansion, improved infrastructure and regional development. This has been achieved despite the lack of significant mineral deposits or other exploitable resources, a feature which partly accounts for Malawi's severe underdevelopment during the colonial era. Historically, this situation underpinned the country's role in the southern African political economy as a small, peripheral labour reservoir (Pachai, 1978; Vail, 1975, 1977, 1983). With a population estimated at more than 6.8 million in 1984 (just over 5.5 million were recorded in 1977), at an average density of 57.7 per square kilometre, creation of additional income earning opportunities has clearly been crucial. The benefits of development have not, however, been distributed with a high degree of equity. As will emerge below, the peasant sector has continued to be underdeveloped during this period of growth.

By contrast, Mozambique is a large littoral state, comprising vast areas of fertile land and a population of only some 12 million in 1980. However, the mid-1984 estimate of 13.4 million (Africa South of the Sahara, 1987), suggests a fairly rapid rate of increase. With a few exceptions, pressure on land is not a problem, although good farming land is scarce in some regions. The average population density in 1984 was only 16.8 per square kilometre. The majority of the population live in the southern and central provinces, between Maputo and Beira, and in Tete Province, which lies between Malawi and Zimbabwe. Maputo, the capital, has been a major regional port since the last century because of its proximity to South Africa's industrial heartland. Beira fulfils a similar role in relation to Zimbabwe's main urban centres. These centres and the export of migrant labour apart, the settler economy was based on estate or *latifundia* agriculture and some mining (Vail and White, 1980). Both activities were, however, limited in extent and large parts of the country remained under virgin bush or peasant cultivation. Territorial integration is poor, with incomplete and grossly inadequate infrastructural links.

A bitter liberation war by FRELIMO preceded independence, which was precipitated by the collapse of Portugal's fascist regime in April 1974. The vast majority of the settler population, who had dominated and controlled the economy, fled, frequently sabotaging facilities. Together with the virtual absence of even semi-skilled indigenes, this precipitated economic chaos. Partly as a result, the new government's nationalisation and state control of the economy as part of an attempted socialist transformation have failed, prompting a re-evaluation and indeed eager acceptance of foreign capital investment since the fourth FRELIMO congress in 1983. However, an escalating and particularly brutal guerrilla insurgency by the South African-backed Mozambique National Resistance (MNR) has brought large-scale destruction, murder, famine and population displacement, undermining most development efforts and wresting effective control of large areas of the country from the state. Even the late President Machel's *rapprochement* with South Africa, embodied in the 1984 Komati Accord, has proved ineffectual (Hanlon, 1986; Isaacman and Isaacman, 1983; Meyns, 1981; Munslow, 1983; Raikes, 1984).

Against this background, the contrasting experiences of the two countries' efforts to integrate tens of thousands of returned migrants are illuminating. Whereas Malawi represented a most unusual case of rapidly increasing domestic labour demand due to labour-intensive economic development, Mozambique was less fortunate and suffered greatly in terms of both unemployment and its balance of payments. Nevertheless, the major impact in both cases was felt in rural areas, especially the peasant sector whence most of the migrants originated. The precise nature of this impact has implications not only for rural development but also urban-rural interaction.

Labour-Intensive Agriculture in Malawi

Malawi's development programme during the late 1960s and 1970s centred on capitalist agriculture, particularly the estate form of production. The chief commodities were cash crops, particularly tobacco, sugar and more recently also tea. World market prices for all three were buoyant over that period. Tobacco and tea production are inherently labour-intensive and, because of the low cost of labour, capital intensive sugar cultivation was eschewed in favour of labour absorption. As Christiansen and Kydd (1983) demonstrate, the labour requirements of estate agriculture were met from two sources, namely expropriation from peasant production, and returning international migrants. The state actively fostered this process through its dominant politico-economic position. This merits some elaboration in view of the important implications for both peasant and commercial agriculture, for rural versus urban livelihood sustaining opportunities, and hence for urban-rural interaction within the country.

Extracting surplus from the peasantry. Notwithstanding some intensive, externally funded peasant sector development projects (Christiansen, 1984; Lele, 1975), Malawi's peasantry have been systematically discriminated against in favour of capitalist agricultural expansion. Three elements in this process can be distinguished (Christiansen and Kydd, 1983). Firstly, the implicit rate of return to peasant cultivation was suppressed by holding the real value of marketing board purchases per capita constant since 1964, a period when real world prices of many crops were rising. The substantial price differentials between marketing board receipts and payments to producers represent direct surplus extraction from the peasant sector. The majority of these proceeds were directed as aid towards the privately owned estates. It should be mentioned, however, that this distortive device is by no means unique to Malawi, having been a key policy instrument for procuring cheap labour throughout British colonial Africa. Whereas Malawian peasant producers were compelled to sell their output to the marketing board, the estates enjoyed free access to world markets, thereby gaining a further advantage over the peasants. The second element of discrimination was that ancillary policy measures induced commercial banks to favour the estates in their allocation of loans, while, thirdly, the state made uncultivated land preferentially available to the estate sector (Christiansen and Kydd, 1983). Bush *et al.* (1986, p.298) argue that the peasantry actually lost land to the estates, and that the landholdings of some 50 per cent of peasant households are now inadequate for subsistence.

The transformation of a predominantly peasant agricultural sector to one characterised by important commercial estate production, is attributed by Christiansen and Kydd (1983) to a combination of these three factors, although they concede that the state may not have anticipated the extent of this effect at the time. Over the 1966-1977 period, full-time male employment in peasant production fell signficantly in absolute terms, although part-time involvement rose. The overall annual increase in peasant employment for both sexes (2 per cent per annum) was lower than Malawi's natural rate of population growth (2.6 per cent per annum) and well below the migration-supplemented rate of increase (2.9 per cent per annum). Moreover, average formal sector wages rose by almost 11 per cent per annum during this period, with part-time female wages increasing by over 22 per cent per annum. As a result of this process and the development of Malawi's new centrally located capital, Lilongwe, the country's pattern of internal population movement, both rural-rural and rural-urban, underwent rapid and significant change (Christiansen, 1984).

Employing returning migrants. Malawi's role as a major labour exporter to the region ended remarkably swiftly. It is possible to argue that the return of migrants from beyond the country's borders was not only *facilitated* by changes contingent upon the expansion of estate agriculture, but possibly also *required* by them. For Banda would probably not otherwise have been able summarily to suspend labour recruitment by South Africa after the 1974 air tragedy. This event may thus have provided a convenient pretext to satisfy domestic labour requirements while professing concern for the welfare of Malawian workers abroad.

South Africa was by no means the only important destination for Malawian labour. In 1972, some 487,000 workers, representing fully 10.3 per cent of the country's *de facto* population, were estimated to be abroad. Roughly 51 per cent of these were in Zimbabwe (then Rhodesia), almost 37 per cent in South Africa, nearly 9 per cent in Zambia and 2 per cent in Tanzania. Some 212,000 migrants, almost half the total, returned home during the five years between 1972 and 1977 as a result of government decree (South Africa), guerrilla wars (Mozambique and Zimbabwe) and anti-Malawian agitation (Zambia). Over the 1966-1977 inter-censal period, about a third of a million migrants are estimated to have returned to Malawi without precipitating a significant increase in *measured* unemployment (Christiansen and Kydd, 1983; my emphasis).

It is important to point out that, although increasing in absolute numbers, less than a quarter of each year's returnees found waged employment. Nevertheless, this proportion is comparable to that for the economically active population as a whole. The majority of each year's returnees, varying from 55 to 73 per cent, were absorbed by the peasant sector (Christiansen and Kydd, 1983; Ettema, 1984). It is not clear to what extent this was their preference or the result of a lack of wage earning opportunities, although the percentage returning to the peasant sector does appear to have risen as the absolute number of returnees increased during the early 1970s. Under the circumstances, *disguised* rural unemployment and underemployment are sure to have increased significantly. Access to land is likely to have been an important criterion in decision-making (however constrained), as is the level of capital accumulated while employed abroad. As will be demonstrated below with respect to Mozambique, international migrants were (and are) by no means a homogeneous group.

There is evidence that many returnees to Malawi engage in seasonal wage employment, thus still necessitating periodic male (and sometimes female) absenteeism. This is almost certainly a consequence of the depressed subsistence conditions and increased land pressure in the peasant sector already referred to (see Bush *et al.*, 1986; Christiansen and Kydd, 1983). A significant proportion of such seasonal or temporary wage labour is undertaken on agricultural estates, although some is also undoubtedly performed in towns and cities. Wages and working conditions on the estates are poor. Periodic rural-urban migration (circulation) and related forms of interaction within Malawi have thus not been eliminated, although they are probably lower than would have been the case without the new agricultural estates. Instead, the country is experiencing higher levels of rural-rural interaction between two unequally favoured modes of agricultural production. Furthermore, some 27,000-30,000 predominantly rural Malawians still find employment in South African mines each year (Table 5.1).

It is clear that Malawi, a small, conservative state in the outer periphery of southern Africa, has reduced its dependence on South Africa (the key objective of SADCC) and Zimbabwe by absorbing over a third of a million former international labour migrants in a remarkably short time without a significant impact on measured domestic unemployment. This has occurred through expansion of large-scale commercial estate production, which drew in peasant labour, to be replaced in turn by a high proportion of the returnees. Less clear, however, is how sustainable the chosen development path will prove, given the degree of ongoing surplus extraction from the peasant sector, which has also experienced growing population pressure and land scarcity. Malawi's economic stagnation during the early 1980s and again since 1986, due to a combination of world market conditions, drought, the impact of war in neighbouring Mozambique and inefficient domestic resource allocation between the estate and peasant sectors, may well be exacerbating the problem to a critical level. In other words, the viability of even sub-subsistence peasant production and reproduction appears under threat. Should this become a reality, pressure on both commercial agricultural land and Malawi's urban areas will increase dramatically, through rapid inmigration of stricken peasants. Some ameliorative measures with respect to the peasant sector were taken in 1983, but their effect has yet to be documented.

Crisis and Stagnation in Mozambique

The export of labour to South Africa, together with earnings from South African freight traffic through Maputo, underwrote Mozambique's colonial economy. The crucial importance of labour migration was two-fold, providing direct employment for over 100,000 workers surplus to domestic requirements, and generating indirect foreign exchange earnings. The latter arose because part of the migrants' wages were remitted to Mozambique in gold valued at the official gold price (US$38 per ounce) when this still existed. The colonial authorities sold it at the far higher prevailing world market prices and appropriated the difference. After 1973, these cosy arrangements collapsed because of the abolition of a fixed official gold price, the great reduction in the number of migrants working in South Africa (see above) and, following independence under a socialist and strongly anti-apartheid FRELIMO government, South Africa's refusal in 1979 to continue paying migrants' remittances in gold at below market prices. Combined with the mass settler flight at independence, South Africa's dramatic, politically-motivated reduction of transit traffic through Maputo, and the new state's vigorous pursuance of inappropriately centralised production, this loss of foreign exchange precipitated economic and food supply crises. The government's initial concentration on state farms, comprising much of the former colonial agricultural sector, was in part a crisis response to maintain output levels, but also reflected an inadequate conception of the need to change colonial food consumption patterns to more appropriate forms (Raikes, 1984). Partial modification of these policies and economic recovery since the early 1980s have, however, been outweighed by the destruction wrought by the MNR insurgency.

The context in which Mozambique has been compelled to seek domestic reincorporation of returning migrants could not have been more difficult, and indeed different from that of Malawi. Perhaps the only common point is that the urban secondary and tertiary sectors were unable to absorb more than a small minority of returnees, and these were generally the better skilled. State bureaucracies aside, Mozambique's industrial base is even more limited than that of Malawi, and for the reasons indicated above, employment in it has not expanded significantly in overall terms since independence. At least non-agricultural employment in Malawi grew consistently until the recession of the early 1980s.

Under such circumstances, returnees clearly had to seek absorption in the rural economy. More specifically, given the limited scale of commercial agriculture and poor short term prospects for its expansion, this burden fell on the peasant sector, particularly in southern Mozambique whence most of the migrants originated. Furthermore, as Raikes (1984) shows, the peasant sector was not bolstered by state policy during the first seven or eight years of independence. First's (1983) landmark study, part of a major project to examine the problems and derive solutions, provides fascinating insights into the history and impact of labour migration to South Africa in both structural and human terms. The following paragraphs draw primarily on her study.

Conditions for the peasantry in Inhambane Province, one of the main sending areas, are bad, and migrant flows have varied historically according to the extent of agricultural crisis. As elsewhere in southern Africa, colonial capitalist penetration had transformed the rural economy, usurping the highest quality and best watered land for settler occupation, and undermining peasant production by coerced internal (*chibalo*) and international labour migration and by distortion of indigenous land allocation mechanisms. Moreover, unlike the settler *latifundarios*, peasants had no access to tied labour. In some areas, a relatively open land frontier permitted expansion of holdings by bush clearance. However, the size of household workforce and access to the necessary implements determined the ability to exploit the opportunity. Overall, land holdings were small and inequitably distributed, with 35 per cent of peasant households having access to less than 1 hectare (totalling only 12 per cent of occupied land) and only 7 per cent occupying more than 4 hectares (23 per cent of the land). The mean holding size of 2 hectares was, however, exceeded in only two of the country's ten provinces. Taken together with the fact that there was no absolute land shortage (only scarcity of productive land), and consequently, no class of landless peasants, this suggests that Inhambane was *relatively* well off. During the 1970s, peasant production increased somewhat. This is partly attributable to changing state policies after independence, coupled with spontaneous or politically organised peasant occupation of some former settler holdings. The most important factor, however, was an enhanced ability to invest capital accumulated from increased real wages by migrants to South Africa. Although virtually all peasant households had sent workers to the mines, they were not all reliant on, and affected by, miners' wages to the same extent. Limited social differentiation into clearly discernible poor and middle peasantries was arising ".... from these groups' differential reliance on wage labour, and on the different wage and skill levels within the mining industry. The stratum of middle peasants is able, after a period of wage work on the mines, to sustain a certain level of agricultural production. By contrast, the poor peasants cannot withdraw from the contract cycle; their agricultural base is too poor to sustain their families without contract renewals, and they regard themselves as unemployed when not on contract" (First, 1983, p.185).

This raises three important issues in the present context. Firstly, the existence of a significant agricultural surplus-producing middle peasantry effectively "disposes of any notion that the *invariable* action of mining capital in these peasant labour societies is to reduce the social formation to a mere reserve army of labour. At the same time these peasants cannot be described as independent petty commodity producers, since acquisition of means of production and, at times, even of sources of reproduction, derives from mine wage labour" (First, 1983, p.185; my emphasis).

What we have is therefore an ongoing situation of incomplete and impermanent urban proletarianisation, which serves capitalist interests provided that the labour extraction does not precipitate a total collapse of rural peasant subsistence production. This gives rise to a large and increasingly differentiated class of people who are *both* workers *and* peasants, or perhaps more accurately, *neither* settled workers *nor* full-time peasants. By definition, then, they are forced to retain both urban and rural footholds, linked by oscillating labour migration and wage and commodity remittances, in order to survive. The degree of differentiation depends on the precise balance of individual households' rural-urban relationships and their degree of choice or flexibility (Bush and Cliffe, 1984; First, 1983; Simon, 1985). Of course, just like the political economy, this class transcends national boundaries, linking far flung rural peasantries with the subcontinent's urban-industrial cores. As hinted earlier, the collapse of peasant production and even social reproduction is now a real problem in densely populated parts of rural southern Africa.

The second important issue follows directly from this point. The ending of migration and the provision of alternative employment opportunities will depend on the transformation of relations of production in the rural areas from which that labour has been drawn. This not only requires rural development as such, but a particular type of transformation. Both Mozambique's existing state farm sector, and experiences elsewhere with export-oriented cash crop strategies, suffer from numerous problems and limitations. At the same time, Mozambique's peasant sector clearly needs to break with its colonial legacy of dependence on at least seasonal wage labour. Hence, First (1983) saw co-operative peasant production, perhaps in conjunction with collective villages (see Rogerson, 1981), as the most viable alternative for rural social(ist) transformation. However, she made no specific policy proposals beyond pointing to some of the problems faced elsewhere. Following acceptance of the failings of earlier policies, FRELIMO's fourth congress in 1983 resolved not to expand the state farm sector but to concentrate instead on co-operatives, peasants and private farmers. At that stage, though, no clear idea of concrete policy measures was articulated (Raikes, 1984). In order to explore the practical implications of such policies in more detail, recent experience in Zimbabwe is analysed in the next major section of the chapter.

The third issue raised by increasing peasant differentiation is that existing sexual divisions of labour and rural relations of production will be affected. The extent of change will depend, not only on the actual differentiation but also on the long-term impact of migrant labour on household structure and marital relationships, together with officially encouraged moves towards women's liberation (see Urdang, 1984). What this implies for returning migrants, for the concept of the family, urban-rural relations, the future of the peasantry and state agricultural policy, can as yet be mere conjecture.

As a final point on Mozambique, however, the limited evidence available suggests that, despite the country's avowed objectives of socialist transformation and reducing dependence on South Africa, it has ultimately been far less successful in absorbing returnee migrants in line with the latter policy than conservative Malawi. During the early 1980s, guerrilla destruction in rural areas, agricultural stagnation and pressure on good peasant land, induced increasing population displacement towards the main cities, particularly Maputo. Despite the lack of significant secondary sector expansion, Maputo's population is estimated to have doubled in the first eight years of independence, and to have reached over 900,000 by mid-1984 (*Guardian*, 5th July 1983; Africa South of the Sahara, 1987). Irregular housing areas now provide shelter for over half the city's population. It is unclear to what extent returning migrants from South Africa, together with their families, have contributed to this growth. With rampant unemployment, chronic food supply problems resulting both from domestic underproduction and erratic foreign aid deliveries, and increasingly serious MNR encroachment on the capital, a rigid system of food ration cards and residence permits was hastily introduced. The urban escape option for remaining rural dwellers has thus been foreclosed, while government services and aid penetrate beyond the immediate urban hinterlands only intermittently during times of MNR activity. By the end of 1986, for example, no fewer than 595 (30 per cent) of the country's 1,921 primary health posts and clinics had been destroyed, looted or forced to close (*Guardian*, 15th January 1988). What has become virtually a survival crisis for the state is thus seriously affecting both urban and rural populations, restricting urban-rural mobility and currently actually threatening the lives of up to one third of Mozambique's population.

Agrarian reform and population mobility in Zimbabwe

Despite its rhetoric of radical social transformation to integrate the country spatially, politically and economically, and to provide social ownership of the means of production and reproduction, the government of independent Zimbabwe has pursued a more pragmatic and arguably non-socialist policy of 'redistribution through growth with equity'. Since just under 75 per cent of the official 1982 population of 7.5 million are classified as rural, agriculture is clearly a key economic sector. The highly fragmented communal lands (former tribal trust lands) of the space economy's periphery, suffer problems typical of southern African reserves, with severe land pressure and degradation in certain regions. The national average population density in 1984 was, however, only 20.9 per square kilometre (Africa South of the Sahara, 1987). Central to state policy for forging a single, integrated agricultural sector is comprehensive agrarian reform. This would inevitably entail significant population resettlement. Indeed, the Riddell Commission (Government of Zimbabwe, 1981) argued that 455,000 of the approximately 780,000 families in the communal lands would be thus affected. Some resettlement on former white commercial farms has taken place (30,800 families by mid-1984), but with a national population growth rate of over 3.1 per cent per annum, the numbers requiring resettlement because of landlessness or overcrowding are increasing rapidly, however successful the actual resettlement. Moreover, the position of peasant women relative to men has not improved. They have been marginalised in subsistence labour without control over the land they till, as a result of their menfolk's absence performing migrant labour. Yet, despite the state's commitment to sexual equality and women's liberation, the resettlement eligibility criteria have failed to increase their rights over land and other resources (Bratton, 1986; Bush and Cliffe, 1984; Gordon, 1984; Jacobs, 1984; Kinsey, 1982; Simon, 1985, 1986; Weiner *et al.*, 1985; Whitlow, 1980; Zinyama, 1982, 1986a, 1986b). Little other rural land outside the national parks and game reserves is available for resettlement.

The Riddell Commission's major proposal for relieving land pressure and facilitating agrarian reform was urbanisation and, in particular, the reunification in urban areas of the approximately 235,000 families it estimated to have been separated by migrant labour. This was essentially a prescription for ending migrant labour by dividing the worker-peasant class into a permanently settled urban proletariat on the one hand, and a full-time peasantry on the other. Nevertheless, a small proportion, comprising especially female headed households, would be overlooked. The new government had a strong commitment to eliminating migrant labour which they saw as symbolising the iniquities of colonial capitalism. However, the Riddell proposals differ little in essence from earlier 'liberal' plans predicated on concern with the social evils of migrancy rather than the social relations of production. In this light, state resettlement policy is inadequate for several reasons, not least its implications for women (see Bush and Cliffe, 1984; Jacobs, 1984; Simon 1985).

For present purposes, I shall focus on the urbanisation aspects of the agrarian reform and resettlement policies. According to the Riddell Commission, the 235,000 split families would be encouraged to move voluntarily by higher urban minimum wages and improvements in housing conditions and social services, while the remaining 185,000 families requiring resettlement from the communal lands, and not catered for in rural areas, would gradually be absorbed by expanding industrialisation.

These proposals can be criticised on at least three grounds. Firstly, recession from 1982 onwards caused stagnation and even a decline in industrial employment (Riddell, 1984), although there has subsequently been a limited recovery. Nevertheless, there can be little realistic scope in the foreseeable future for absorbing labour on the scale envisaged by Riddell. Secondly, if the movement of migrants' dependents were to be encouraged by consciously increasing the quality of life differentials between communal lands and urban areas, it seems certain that other people, superfluous to urban labour requirements, would also be attracted. With an average family size of six, merely reuniting the split families would add 1,175,000 to the country's urban population. This would occur at a time when the major cities are already growing rapidly, unemployment is rising, and the state is practising restrictive urban policies such as demolishing squatments, suppressing the informal sector and sending unemployed people to rural resettlement schemes. Thirdly, and in consequence, implementation of these Riddell Commission proposals would aggravate already serious urban problems of inadequate shelter, employment and facilities. Furthermore, no consideration was given to how the two groups of potential migrants might be distinguished and those people not able to be absorbed in urban areas dissuaded from moving (Drakakis-Smith, 1987; Simon, 1985, 1986).

Two main lessons emerge. Firstly, rural and urban policy formulation and planning cannot meaningfully be undertaken in isolation. Together with any interstices between them, rural and urban areas (or sectors) constitute a complex and dynamic national, and in some cases international, system. This is the appropriate spatial framework for analysis and within which rural-urban interaction takes place, its form and nature responding to changing conditions. The existing roles of, and policy implications for, men, women and different household structures must be taken into account. The necessity for a systems approach to migration studies was recognised long ago (Mabogunje, 1970), but as we have seen, this frequently does not occur in practice. Secondly, whatever the ideological rationale, efforts to reduce inequitable rural-urban interaction, most particularly human migration, cannot succeed without coercion unless the underlying structural disparities are ameliorated. Attempts to eliminate migrant labour or to absorb returning international migrants by creation of full-time peasants and permanent urban proletarians will fail unless conditions adequate for both production and social reproduction exist in rural and urban areas respectively. At present, survival for most worker-peasants in southern Africa depends on maintaining footholds, however precarious, in *both* rural and urban areas.

Conclusion

Specific conclusions were drawn in relation to each national case study and will not be repeated here. All that remains is to highlight some key issues. This chapter has sought not only to provide a synopsis of the recent literature on migration and rural-urban interaction in southern Africa, but also to highlight the utility of a political economy perspective in analysing the complexities of history, structure, process and policy. Fruitful analysis and policy depend crucially on the adoption of an appropriate systemic framework, encompassing the relevant set of rural and urban areas in one or more countries. However, this needs to be balanced with a proper concern for the people involved, something frequently lost in macro-scale structural analyses.

Institutionalised labour migration within and between countries was a crucial mechanism underlying the establishment of an integrated regional political economy characterised by poverty and inequality. South Africa's economic dominance ensured the persistence of large-scale international migration well beyond the decolonisation of most countries in the region. The importance of these flows to, and the impact of their recent reduction on, national development in countries as different as South Africa, Malawi, Mozambique and Zimbabwe has been clearly demonstrated.

Outcomes are determined not by physiographic and demographic factors alone but, crucially, by how these interact with the respective modes of production, regional and national political economies, state ideologies, and interstate relations. Hence, perhaps somewhat surprisingly, Malawi has been notably more successful in the short to medium term than Mozambique in absorbing its returning migrants. What remains to be seen, however, is whether Malawi's development strategy can be sustained or whether recent allocative distortions will precipitate a longer term crisis. Conversely, if the war in Mozambique can be ended, long term transformation of social relations of production may prove both possible and necessary, as on the one hand the land base and other resources are adequate and on the other, incomplete absorption of returned migrants coupled with existing inequalities and inefficiencies may provide powerful political stimuli. Just how far down the road of socialist transformation Zimbabwe will be prepared to go now that the one-party state is a reality, remains to be seen. Despite dramatic progress in education and health, for example, many of its urban and agrarian reform policies have thus far differed insufficiently from pre-independence practice to achieve an adequate restructuring of productive relations and the elimination of labour migration. Whichever path it ultimately follows, Zimbabwe certainly has the potential to develop into a leader on the subcontinent second only to South Africa. As Parson (1984) points out, however, many of the region's peripheral countries face profound development problems by virtue of their location and mode of insertion into the world economy. Their difficulties in reabsorbing migrant labour previously exported to South Africa, in many ways symbolic of this wider impasse, are thus compounded.

Rural-Urban Interaction in Southern Africa

REFERENCES

Africa South of the Sahara (1987), Europa, London

Amin, S. (1972) 'Underdevelopment and dependence in black Africa: origins and contemporary forms', *Journal of Modern African Studies* 10, 503-24

Bell, M. (1986) *Contemporary Africa*, Longman, London

Beinart, W. (1982) *The Political Economy of Pondoland 1860-1930*, Cambridge University Press, Cambridge

Bley, H. (1971) *South West Africa under German Rule*, Heinemann, London

Bratton, M. (1986) 'Farmer organizations and food production in Zimbabwe', *World Development* 14, 367-84

Bundy, C. (1979) *The Rise and Fall of the South African Peasantry*, Heinemann, London

Bush, R. and Cliffe, L. (1984) 'Agrarian policy in migrant labour societies: reform or transformation in Zimbabwe?', *Review of African Political Economy* 29, 77-94

Bush, R., Cliffe, L. and Jansen, V. (1986) 'The crisis in the reproduction of migrant labour in Southern Africa', in: Lawrence, P. (ed.) *World Recession and the Food Crisis in Africa*, James Currey, London, 283-99

Christiansen, R. (1984) 'The pattern of internal migration in response to structural change in the economy of Malawi 1966-77', *Development and Change* 15, 125-51

Christiansen, R.E. and Kydd, J.G. (1983) 'The return of Malawian labour from South Africa and Zimbabwe', *Journal of Modern African Studies* 21, 311-26

Cobbe, J. (1986) 'Consequences for Lesotho of changing South African labour demand', *African Affairs* 85, 23-48

Cronje, G. and Cronje, S. (1979) *The Workers of Namibia*, International Defence and Aid Fund, London

Crush, J.S. (1980a) 'The genesis of colonial land policy in Swaziland', *South African Geographical Journal* 62, 73-88

Crush, J.S. (1980b) 'The colonial division of space: the significance of the Swaziland land partition', *International Journal of African Historical Studies* 13, 71-86

Crush, J.S. (1984) 'Uneven labour migration in southern Africa: conceptions and misconceptions', *South African Geographical Journal* 66, 115-32

Crush, J.S. (1986) 'The extrusion of foreign labour from the South African gold mining industry', *Geoforum* 17, 161-72

Drakakis-Smith, D. (1987) 'Urban and regional development in Zimbabwe', in: Forbes, D. and Thrift, N. (eds.) *The Socialist Third World; Urban Development and Territorial Planning*, Blackwell, Oxford

Ettema, W. (1984) 'Small-scale industry in Malawi', *Journal of Modern African Studies* 22, 487-510

Ettinger, S. (1972) 'South Africa's weight restrictions on cattle exports from Bechuanaland 1924-41', *Botswana Notes and Records* 4, 21-33

First, R. (1983) *Black Gold: the Mozambican Miner, Proletarian and Peasant*, Harvester, Sussex and St Martin's, New York

Gordon, D.G. (1984) 'Development strategy in Zimbabwe: assessment and prospects', in: Schatzberg, M.G. (ed.) *The Political Economy of Zimbabwe*, Praeger, New York

Gordon, R.J. (1977) *Mines, Masters and Migrants: life in a Namibian Compound*, Ravan, Johannesburg

Government of Zimbabwe (1981) *Report of the Commission of Inquiry into Incomes, Prices and Conditions of Service*, Government of Zimbabwe, Salisbury

Guardian, (1983) 5th July, London

Guardian, (1988) 15th January, London

Hanlon, J. (1986) *Beggar Your Neighbours: Apartheid Power in southern Africa*, James Currey, London

Isaacman, A. and Isaacman, B. (1983) *Mozambique: From Colonialism to Revolution, 1900-1982*, Gower, Aldershot

Jacobs, S. (1984) 'Women and land resettlement in Zimbabwe', *Review of African Political Economy* 27/28, 33-50

Kinsey, B.H. (1982) 'Forever gained: resettlement and land policy in the context of national development in Zimbabwe', *Africa* 52, 92-113

Legassick, M. (1975) 'South Africa: forced labour, industrialization and racial differentiation', in: Harris, R. (ed.) *The Political Economy of Africa*, Schenkman, Cambridge

Lele, U. (1975) *The Design of Rural Development; Lessons from Africa*, Johns Hopkins University Press (for the World Bank), Baltimore

Lemon, A. (1982) 'Migrant labour and frontier commuters: reorganising South Africa's black labour supply', in: Smith, D.M. (ed.) *Living Under Apartheid*, Allen and Unwin, London, 64-89

Leys, R. (1975) 'South African gold mining in 1974: the gold of migrant labour', *African Affairs* 74, 196-108

Lye, W.F. and Murray, C. (1980) *Transformations on the Highveld: the Tswana and South Sotho*, David Philip, Cape Town

Mabogunje, A.L. (1970) 'Systems approach to a theory of rural-urban migration', *Geographical Analysis* 2, 1-18

Massey, D. (1980) 'The changing political economy of migrant labour in Botswana', *South African Labour Bulletin* 5, 4-26

Meyns, P. (1981) 'Liberation ideology and national development strategy in Mozambique', *Review of African Political Economy* 22, 42-64

Moorsom, R. (1977a) 'Underdevelopment and class formation: the origin of migrant labour in Namibia 1850-1915', in: Adler, T. (ed.) *Perspectives on South Africa*, University of Witwatersrand Press, Johannesburg, 3-50

Moorsom, R. (1977b) 'Underdevelopment, contract labour and worker consciousness in Namibia, 1915-72', *Journal of Southern African Studies* 4, 52-87

Munslow, B. (1983) *Mozambique: The Revolution and its Origins*, Longmans, London

Murray, C. (1981) *Families Divided: the Impact of Migrant Labour in Lesotho*, Cambridge University Press, Cambridge

O'Connor, A. (1983) *The African City*, Hutchinson, London

Pachai, B. (1978) *Land and Politics in Malawi 1875-1975*, Limestone Press, Kingston, Ontario

Palmer, R. (1977) *Land and Racial Domination in Rhodesia*, California University Press, Berkeley

Parson, J. (1984) 'The peasantariat and politics: migration, wage labour and agriculture in Botswana', *Africa Today* 4th quarter, 5-25

Phimister, I. (1977) 'Peasant production and underdevelopment in southern Rhodesia 1890-1914', in: Palmer, R. and Parsons, N. (eds.) *The Roots of Rural Poverty in Central and Southern Africa*, Heinemann, London

Raikes, P. (1984) 'Food policy and production in Mozambique since independence', *Review of African Political Economy* 29, 95-107

Rural-Urban Interaction in Southern Africa

Republic of South Africa (1985) *South Africa 1985: Official Yearbook of the RSA (11 Edition)*, Department of Foreign Affairs, Pretoria

Republic of South Africa (1986) *South Africa 1986: Official Yearbook of the RSA (12 Edition)*, Department of Foreign Affairs, Pretoria

Rich, P.B. (1978) 'Ministering to the white man's needs: the development of urban segregation in South Africa 1913-1923', *African Studies* 37, 177-91

Riddell, J.B. (1981) 'Beyond the description of spatial pattern: the process of proletarianization as a factor in population migration in West Africa', *Progress in Human Geography* 5, 370-92

Riddell, R.C. (1984) 'Zimbabwe: the economy four years after independence', *African Affairs* 83, 463-76

Rogerson, C.M. (1981) 'The Communal villages of Mozambique; an experiment in rural socialist transformation', *Geography* 66, 232-35

Simon, D. (1985) 'Agrarian policy and migration in Zimbabwe and southern Africa: reform or transformation?', *Review of African Political Economy* 34, 82-89

Simon, D. (1986) 'Regional inequality, migration and development: the case of Zimbabwe', *Tijdschrift voor Economische en Sociale Geografie* 77, 7-17

Swindell, K. (1979) 'Labour migration in underdeveloped countries: the case of sub-Saharan Africa', *Progress in Human Geography* 3, 219-59

Taylor, J. (1982) 'Changing patterns of labour supply to the South African gold mines, *Tijdschrift voor Economische en Sociale Geografie* 73, 213-20

Taylor, J. (1986) 'Some consequences of recent reductions in mine labour recruitment in Botswana, *Geography* 71, 34-46

Urdang, S. (1984) 'The last transition? Women and development in Mozambique', *Review of African Political Economy* 27/28, 8-32

Vail, L. (1975) 'The making of an imperial slum: Nyasaland and its railway, 1895-1935', *Journal of African History* 16, 89-112

Vail, L. (1977) 'Railway development and colonial underdevelopment: the Nyasaland case', in: Palmer, R. and Parsons, N. (eds.) *The Roots of Rural Poverty in Central and Southern Africa*, University of California Press, Berkeley

Vail, L. (1983) 'The state and the creation of colonial Malawi's agricultural economy', in: Rotberg, R. (ed.) *Imperialism, Colonialism and Hunger* , D.C. Heath, Lexington, 39-88

Vail, L. and White, L. (1980) *Capitalism and Colonialism in Mozambique*, Heinemann, London

Van Onselen, C. (1976) *Chibaro; African Mine Labour in Southern Rhodesia, 1900-1933*, Pluto, London

Weiner, D., Moyo, S., Munslow, B. and O'Keefe, P. (1985) 'Land use and agricultural productivity in Zimbabwe', *Journal of Modern African Studies* 23, 251-85

Wellings, P.A. (1986) 'Lesotho: crisis and development in the rural sector', *Geoforum* 17, 217-37

Whitlow, J.R. (1980) 'Environmental constraints and population pressures in the tribal areas of Zimbabwe', *Zimbabwe Agricultural Journal* 77 173-81

Wilson, F. (1972) *Labour in the South African Gold Mines 1911-1969*, Cambridge University Press, Cambridge

Wilson, F. (1972) *Migrant Labour in South Africa*, South African Council of Churches and SPROCAS, Johannesburg

Wolpe, H. (1972) 'Capitalism and cheap labour power in South Africa: from segregation to apartheid', *Economy and Society* 1, 425-56

Zinyama, L.M. (1982) 'Post-independence land resettlement in Zimbabwe', *Geography* 67, 149-52

Zinyama, L.M. (1986a) 'Agricultural development policies in the African farming areas of Zimbabwe', *Geography* 71, 105-15

Zinyama, L.M. (1986b) 'Rural household structure, absenteeism and agricultural labour: a case study of two subsistence farming areas in Zimbabwe', *Singapore Journal of Tropical Geography* 7, 163-73

6

THE URBAN-RURAL CONTEXT OF AGRARIAN CHANGE IN THE ARABIAN PENINSULA

TIM UNWIN

During the last two decades the agrarian scene in the Arabian Peninsula has been dramatically transformed. Nowhere is this more apparent than in the Kingdom of Saudi Arabia, where in the decade 1975-1985 according to official government figures the total cultivated area increased from 150,000 to 2,300,000 hectares. This generated considerable increases in arable production, and in the eight years from the 1977-78 harvest to the 1985-86 harvest wheat production in the Kingdom rose from about 3,000 tons to around 2,000,000 tons (Kingdom of Saudi Arabia, Ministry of Agriculture and Water, 1986). This chapter explores the economic and social transformations that have underlain such staggering achievements, and it focuses particular attention on the changing urban-rural relations which have both engendered agricultural change, and yet which have also at the same time been one of the results of such change (Figure 6.1).

The production of food remains the most basic and important economic activity of mankind. As the recent famines in parts of Africa have emphasised only too vividly, without food, mass rural starvation and social chaos ensue. However, it is all too easy to forget that by their very nature all urban settlements can only be sustained by a permanent supply of food. As urbanisation continues, there is an ever growing requirement to increase the flow of food to sustain a rising urban population. Analysis of food flows must therefore lie at the heart of any understanding of urban-rural relations, and this chapter explores the changing patterns of such flows in the Arabian peninsula over the last thirty years. At the beginning it is important to stress a

Figure 6.1: The Arabian peninsula

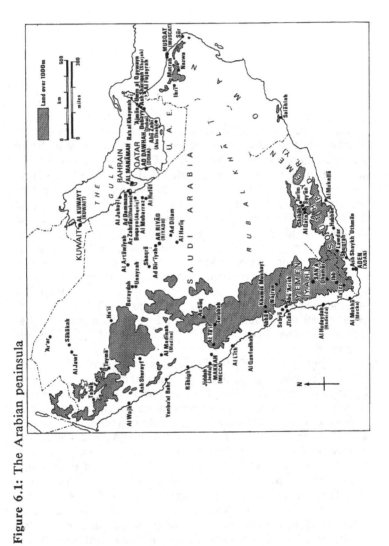

fundamental distinction between two types of food flow associated with urbanisation. On the one hand there is urbanisation which is sustained by increased agricultural production in rural areas within the country where that urbanisation is taking place, and on the other there is urbanisation that is sustained essentially by the importation of food supplies from other countries. The former situation is well exemplified in the context of urbanisation and industrialisation in 19th century England, when English agriculture was on the whole able to raise its production broadly in step with population increases (Wrigley and Schofield, 1981). Although England did import much food during this period, the dominant flow of food was from the countryside to the towns and industrial regions. In contrast, during the 1970s the rapid urbanisation that was taking place in the Arabian peninsula was fuelled primarily by imports of food from overseas. The Arabian peninsula is thus a classic instance of a situation where urbanisation has far outstripped the capacity of the land to support it, and where changes in agriculture have followed on from preceding urban and industrial change.

This chapter focuses attention on the factors behind these changes, and it begins with a broad summary of the demographic situation and the balance between the urban and rural economies. This is followed by an evaluation of the agriculture and food flows of the region prior to the exploitation of its oil resources, and then the agricultural changes that have recently taken place are analysed at greater length in the context of the development of the oil industry which generated the rapid urbanisation of the 1970s.

Urbanisation in the Arabian Peninsula: a statistical overview

Obtaining accurate statistical information on the countries of the Arabian peninsula, particularly concerning its demography and agricultural production, is even today extremely difficult. However, prior to the 1970s it is well nigh impossible. As Bonine (1980, p.226) has emphasised when writing on the region, "statistics on city growth and urbanization are inadequate, inaccessible, or simply nonexistent".

171

Table 6.1 nevertheless brings together a number of estimates of the urban population density of countries in the region between 1960 and 1984. While no great reliance should be placed on the accuracy of these figures, they do highlight three broad features summarising the nature of urbanisation in the region. The first is that over the last twenty years, regardless of the initial percentage of the population that was urban, all of the countries have undergone very rapid and extensive urbanisation. Even in Oman, which has the lowest density of urban population, the percentage of the population that is classified as urban has approximately doubled over a twenty year period. The second feature to note is that the *total* population of the region has also increased dramatically, and that most of this growth has been accounted for by increases in the urban population. This situation has been well summarised by Bonine (1980, p.260) when he commented that "the Arab shaikhdoms of the lower Persian Gulf have undergone some of the most dramatic growth that has ever propelled a society into the twentieth century. The recent growth rates of these countries and cities ... are the highest in the world". Much of this population increase is the result of the immigration of foreign workers who have supplied the much needed labour for the rapid industrial development that has taken place (Birks and Sinclair, 1980). The implications of these two factors for agriculture are starkly apparent, and amount to a rapidly increasing demand for food at a time when the percentage of population in rural areas has been declining. The third feature to note from Table 6.1 is that there are great differences between the countries in the region, both in terms of their demographic structure in the early 1960s and also in terms of the changes that have taken place during the last two decades. Generally speaking, it is the oil-rich states, such as Bahrain, Kuwait, Qatar, Saudi Arabia and the United Arab Emriates that have encountered the greatest amount of urbanisation and total increases in population, whereas in contrast, Oman and the Yemen Arab Republic still retain considerable rural populations. Nevertheless, the nature of urban settlement prior to the discovery of oil is also significant for an understanding of the changes that have taken place over the last twenty years, and some attention

Table 6.1: Urban population percentages in the Arabian Peninsula, 1960-1984

	Percentage Urban Population				Total Population (millions)	
	1960	1970	1976	1984	1972	1984
Bahrain	-	64.2	77.7	81	0.2	0.4
Kuwait	69	56.3	88.4	90	0.8	1.6
Oman	-	4.9	5.3	8	0.7	1.0
People's Democratic Republic of Yemen	20	28.7	-	38	1.4	2.1
Qatar	-	68.3	88.7	87	0.1	0.3
Saudi Arabia	12	23.6	20.8	70	8.2	10.8
United Arab Emirates	-	51.9	84.0	81	0.2	1.5
Yemen Arab Republic	4	5.8	-	38	6.1	5.9

Source: Clarke (1981, p.158); Beaumont, Blake and Wagstaff (1976, p.177 and p.205); Kubursi (1984, p.37); Drysdale and Blake (1985, p.15-16)

should therefore briefly be paid to the factors underlying the 'traditional' urban pattern of the peninsula.

Five broad features can be seen to have influenced the location and nature of the towns that existed in the region at the beginning of the 20th century: trade, religion, fishing, pearling, and so-called 'piracy' (Unwin, 1982). Taken together, these factors gave rise to an urban distribution that was primarily coastal in nature. In origin most of the inland towns were based around oases, or lay in the more fertile valleys of the south-west. Riyadh developed into a major town only after it became the Saudi capital during the 19th century, and even then in 1919 it is estimated only to have had a population of around 19,000 (Riyadh Municpality, 1986). Mecca and Medina owed their prominence to their role as pilgrimage centres, and it was only the towns of Yemen that maintained any degree of urban vitality based primarily on agricultural production. Elsewhere, the dominant coastal distribution of the towns owed much to trade. As Yapp (1980, p.47) has observed, in the 19th century "the prosperity of the people of the Gulf depended upon trade and the pearl fisheries". It was this coincidence of pearl fishing and trade, associated with the possibility of obtaining a food supply from fishing that gave rise to the urban settlements of Kuwait, Bahrain and Dubai along the southern shores of the Gulf, and to an extent this therefore explains their relatively high levels of urbanisation in the mid-20th century (Table 6.1). Trade too, explains the importance of such towns as Jeddah, Aden, Mukalla, Salahlah and Muscat. However, to these factors must also be added what the British colonial forces in the region called 'piracy', which was rife in the Gulf during the 18th and 19th centuries (Sweet, 1964). This gave rise to the numerous forts scattered along the coasts of the peninsula, a number of which provided foci for small incipient urban settlements.

Agriculture before oil

The traditional image of the Arabian peninsula is one which sees the population being divided into two, the settled *hadar* and the nomadic bedouin. This dichotomy is reflected in different lifestyles and economies, with the village population being seen as cultivating date gardens while the nomadic bedouin herded their camels across the barren desert in a life-long search for water and pasture. Such an image is vividly brought to mind by Yapp (1980, p.43) in his description of the Gulf region in the 18th century: "its agriculture was meager, the region lacked adequate and reliable supplies of good water, and the ingenuity of its *falajs* or *qanats* (underground irrigation channels) was a palliative only. Cereal production was very rare; fruit, nuts, and vegetables were grown in a few favored locations, and the principal crop - dates - was confined to a select number of places - Buraimi, Liwa, Qatif, Hasa on the Arabian shore, along the Minab river on the Iranian side, and especially on the Batina coast of Oman and around the Shatt al-'Arab at the head of the Gulf. The larger islands, Bahrain and Qishm, were also cultivated. Pastoralism was a feature of the economy, and its products - camels, horses, sheep and goats - were exchanged for the products of cultivation. But the area as a whole required imports of foodstuffs, especially rice, from India". This depressing view, based as it is on a twofold distinction between a settled agricultural population and a nomadic pastoral one, ignores the diversity and skill that had been achieved in agriculture in the region prior to extensive European influence therein. It does nevertheless also illustrate two important features of the agrarian economy: the critical nature of *exchange*, and the need to *import* food from elsewhere.

The written evidence concerning the agriculture of the region in the 19th century is sparse, and in reconstructing an accurate impression of the crops grown and the areas under cultivation it is necessary to have recourse to the writings of early European travellers to Arabia, such as the Bents, Burchardt, Fontanier, Neibuhr, Palgrave, Philby and Thomas. Above all, though, a wealth of information on the economy of the area can be gleaned from the pages of Lorimer's (1908-1915) *Gazetteer of the Persian Gulf, Oman and Central Arabia* (Unwin, 1983).

Three types of exchange were crucial for the traditional economy of Arabia at the end of the last century: exchange between nomads and the settled population; exchange between the fishing and the agrarian communities; and the exchange of imports for exports. Looking at each of these linkages in turn, it is evident that the nomadic bedouin, frequently on the move, required grain, dates and other food from the settled village agricultural population. In turn, the nomads provided the sedentary population with hides, meat and to a lesser extent dairy produce. This simple distinction, though, is not in practice as clear cut as might be imagined, since particularly in the Liwa region of western Abu Dhabi the bedouin spent at least part of the year cultivating their date gardens, and many undertook some pearling. Likewise, it is abundantly evident that the villagers of Arabia also kept large numbers of sheep and goats. Turning to the exchange between fishing and agrarian communities, it is apparent that in many parts of the region fishing was of far more importance than the arable economy. Around all the coasts of the Arabian peninsula village communities existed on food and protein supplied largely from fish. Dried fish was also exchanged by these communities for the products of agriculture, thus forming an important linkage between these two sectors of the community. Where cultivation was limited a close interrelationship was sometimes established between fishing and livestock production. Thus Lorimer (1970, p.241, p.1,186) noted during the early years of the 20th century that cattle in Bahrain were fed, amongst other things, on dates, dried fish and old bones, and that at Muscat 30 cattle and 200 sheep were kept which were fed mainly on fish. The third aspect of exchange in the region was that between imports and exports. In Bahrain, Kuwait, Qatar and much of what is now the United Arab Emirates pearling dominated productive economic activity (Bowen, 1951), and, while there was some limited agricultural production, much of the food other than fish, was obtained from imports, exchanged for the wealth generated by pearls.

It is difficult to estimate to what extent the communities of the Arabian peninsula were self-sufficient in terms of their food production. In the south-west, in much of the Yemens and parts of Asir, it seems likely that little agrarian produce was imported. Likewise, many villagers elsewhere in the region, probably produced most of their own requirements. However, there was little if any agricultural surplus, and the majority of the towns must have relied for their food other than fish primarily on imports. At the beginning of the 20th century Lorimer (1970, p.1,185-86) thus recorded that

"Masqat has no natural rssources (*sic*) or amenities. Food and firewood are all imported, with the exception of the trifling yield of a few date trees and of some small market gardens in Wadi-al-Kabir and of the excellent and abundant fish which a fleet of boats belonging to the town catch outside the entrance of the harbour every morning when the weather permits". Muscat's imports of rice came mainly from Calcutta, and wheat was imported from Karachi, Persia and Iraq. These patterns of trade, although no doubt partly influenced by the British colonial presence in the region, must have reflected very ancient flows of commerce and food. Lorimer (1970, p.1,437-39) also noted that the Trucial Oman, now the United Arab Emirates, imported considerable amounts of grain and pulses from India and Persia. However, a few small towns in the region were more fortunate, and were apparently able to sustain themselves primarily from local production. Thus Lorimer (1970, p.241) noted that the supply of vegetables on Bahrain was sufficient for the population of Manama and Muharraq and that none was imported. The overall picture to be gained from this is that on the whole urban life in the Arabian Peninsula was sustained not by flows of food from the rural areas but rather by imports from overseas.

In order to evaluate what have been the influences of the recent rapid urbanisation on agriculture in the area, it is also necessary to establish precisely what were the main types of crop cultivated, where they were grown, and what methods of cultivation were used in the past.

Agricultural production in Arabia has been limited by both physical and social constraints (Joffe, 1985), but it is the absolute nature of the hostile physical environment that was of prime importance in determining the distribution of agriculture in the region prior to the discovery of oil. As Bowen-Jones and Dutton (1983, p.2) have commented, "underlying all else, is the stark fact that the pursuit of agriculture in Arabia entails a human preparedness either to adapt almost every aspect of social and economic life to the seasonal vagaries of an unmodified arid environment - as with the bedu nomadic pastoralists - or to pay the price, in a variety of ways, of drastically modifying the environment so as to create ecologically fragile conditions reasonably suitable for farming - as with irrigated cultivation". It is not so much the temperature that is the critical environmental variable, but rather the lack of water and the limited soils suitable for cultivation. Rainfed agriculture in Arabia is generally only possible when rainfall levels reach over 200mm per year, and as Table 6.2 indicates this has severely limited the possible amount of land suitable for such agriculture. Only in Oman, Saudi Arabia and the Yemen Arab Republic does rainfall permit any significant amount of agriculture, and even in these countries the rugged mountainous environment prevents most of this land from being utilised for agriculture. The use of ephermeral streams for irrigation did, though, permit sizeable communities to be supported by agriculture in a few favoured areas, as in the Ma'rib area of the Yemen Arab Republic, in the Wadi Hadhramaut region of the People's Democratic Republic of Yemen, in Jizan and Asir in Saudi Arabia, and in parts of the Jebel Akhdar of Oman and the Hajar range in the east of the United Arab Emirates. Elsewhere intricate systems of underground irrigation canals and wells were necessary to enable small areas of suitable soil near the main mountain ranges to be cultivated, and scattered oases also provided limited land for agriculture.

Turning to the actual crops produced, a detailed reading of Lorimer's (1970) 1908-1915 Gazetteer enables us to reconstruct a comprehensive picture of the wealth of crops which it was possible to grow in these areas at the beginning of this century. Looking at Bahrain, Oman, Qatar and the United Arab Emirates, Table 6.3 illustrates the great diversity of crops, particularly of fruit, that was then cultivated, and although this no doubt only presents a limited impression of the diversity of agriculture it is sufficient to suggest that agriculture at this time was much more healthy than some authors, such as Yapp (1980), might give it credit. The extent of commercial and fruit crop production also indicates that this can not all have been for local consumption, and reinforces the argument that some such produce was sold in local coastal towns. The bulk of the cereals required therein must, though, have been imported, primarily from elsewhere in Asia.

In terms of methods of cultivation, the predominant form of landholding was the small family farm. As Bowen-Jones and Dutton (1983, p.5) have emphasised, "for the most part

Table 6.2: Environmental constraints on agriculture in the Arabian Peninsula

	Land Area ('000 hectares)	Area with >200mm mean annual rainfall ('000 hectares)	Land with >200mm rainfall as a percentage of total land area	Percentage of area with >200mm rainfall where terrain does not limit arable land
Bahrain	62	0	0	–
Kuwait	2,072	0	0	–
Oman	21,242	4,460	12	1
People's Democratic Republic of Yemen	28,768	0	0	–
Qatar	2,201	0	0	–
Saudi Arabia	154,600	5,000	3	20
United Arab Emirates	8,360	0	0	–
Yemen Arab Republic	19,500	10,179	52	10

Source: derived from Allan, J.A. (1981)

Table 6.3: Crops and livestock in Bahrain, Qman, Qatar and the United Arab Emirates derived from Lorimer's 1908-1915 Gazetteer of the Persian Gulf, 'Oman and Central Arabia'

FRUIT

Almonds	Dates	Sweet Limes	Oranges	Walnuts
Apricots	Figs	Mangoes	Papaya	
Bananas	Grapes	Musk Melons	Peaches	
Citrons	Guavas	Water Melons	Plums	
Coconuts	Lemons	Mulberries	Pomegranates	
Custard Apples	Sour Limes	Olives	Quinces	

CEREALS

Bajra	Jowar	Millet
Barley	Maize	Wheat

VEGETABLES

Aubergine	Carrots	Leeks	Pulses	Sesame
Beans	Cucumbers	Onions	Radishes	Sweet Potato
Bindis	Garlic	Peas	Red Peppers	

COMMERCIAL AND FODDER CROPS

Bastard Saffron	Indigo	Tobacco
Cotton	Lucerne	Turmeric
Henna	Sugar-cane	

LIVESTOCK

Camels	Dogs	Goats	Poultry
Cattle	Donkeys	Horses	Sheep

Source: derived from Unwin (1983a)

settled rural communities were based on individual family farmholdings, the small size of which was largely determined by the enormous physical difficulty of lifting by man and animal the very large volumes of water required to sustain plants, livestock and their owners under the prevailing climatic conditions". Joffe (1985) has reinforced this point, noting that inheritance patterns resulting from Islamic and customary legal practice were an important factor leading to increased fragmentation of holdings in Saudi Arabia where more than 50 per cent of all agricultural holdings are less than 1 hectare in size. In mountainous areas, where there was limited flat available space for agriculture, extensive systems of terracing were developed, as at Jebel Akhdar, where Lorimer (1970, p.1,743) recorded that "cultivation extends for about 1,000 feet down the hillside upon ledges, only 10 or 12 feet wide, which are frequently revetted. These 'pensile gardens', as they have been called, contain apricots, grapes, figs, pomegranates and grain; the grain is sometimes sown intermixed with leguminous plants and even with melons". Early writers did not comment at length on the actual methods of cultivation used on these holdings, but Lorimer (1970) observed the use of ploughs, hoes and spades, and noted that systems of rotations were in use, with stubble being ploughed in to improve the quality of the limited soils. The critical feature of agriculture was, though, the provision of water, and a great variety of techniques were used to bring water to the agricultural holdings. Wilkinson (1977) has thus observed that the water resources of Oman were traditionally exploited by three techniques: wells, often as much as 45 metres deep; *ghayl aflaj*, where the flow of water in wadi bottoms is diverted for agricultural use; and *qanat aflaj*, which are underground canals dug back into mountain sides to reach underground sources of water. Associated with each irrigation network there was a complex system of water shareholding, known as *dawran* in Oman, which was broadly based on the principle that the original shareholders in the construction of a *falaj* would each receive a similar share of the water.

In contrast to the settled agricultural communities, the way of life of the nomadic bedouin was very different, with a range of different types of nomadism being devised to ensure that sufficient food could be found for the livestock throughout the year. Pastureland was traditionally always tribally controlled, either as *dira* land belonging to pastoralists, or in much smaller amounts as *hima* land belonging to settled communities (Joffe, 1985; Raswan, 1930). However, within the pastureland many different nomadic customs were practised. Janzen (1986, p.53) in studying the bedouins of Dhofar has thus suggested that they can usefully be divided into three main types "on the strength of the differences in the physical lay-out of the environment and the long distances over which they travel, the nature and scope of their stock-breeding and other economic activities" and on "the arrangement of their living quarters and accommodation". To illustrate the complexity of pastoral nomadism in this region he goes on to distinguish between long-range and short-range nomadism, between horizontal and vertical nomadism, and between camel, goat and cattle breeding nomads. Perhaps the most important feature of pastoral nomadism, as it was practised in the Arabian peninsula, was that it was an ecologically *efficient* system that enabled a population to be supported in the long term in an inherently hostile physical environment. A similar balance between water supply and demand had been reached by the settled population and, it is changes in this balance that have been the most significant result of the oil industrialisation and urbanisation in the region to which attention now turns.

Oil and industrialisation

The earliest oil concessions in the Arabian Peninsula were granted in the early 1930s, and production of oil first began in Bahrain in 1932. However, the amount of oil produced in the region was negligible until after the Second World War, when new discoveries were made in Kuwait and then in Saudi Arabia. Prior to the 1970s most production was in the hands of the British or American companies who had gained concessions from the various governments of the peninsula, but during the 1960s these governments began to express a growing desire for greater participation in the existing concessions (McLachlan, 1980). This followed the establishment of the Organization of the Petroleum Exporting

Countries (OPEC) in 1960 by Iran, Iraq, Kuwait, Saudi Arabia and Venezuela. Qatar joined OPEC a year later in 1961, and in 1967 Abu Dhabi also became a member. Nevertheless, it was not until the 1970s that major changes took place in the oil industry. Another important factor in the establishment of greater local control over the oil industry was the acquisition of independence from Britain of the People's Democratic Republic of Yemen in 1967, and then Bahrain, Qatar and the United Arab Emirates in 1971. Until then there had undoubtedly been some industrial progress in the region, with oil refineries having been established in Bahrain in 1936, in Kuwait in 1946 and in Saudi Arabia soon afterwards, but independence acted as a major stimulus to economic change. The oil price rises of 1973, when the official price of Saudi Arabian Light rose from U.S.$ 2.75 per barrel in July 1973 to U.S.$ 10.84 per barrel in January 1974, then provided the wherewithall to finance these changes. From December 1972, under the General Agreement on Participation, the OPEC countries took an immediate 25 per cent stake in the concessionaire companies, and since then most countries in the region have achieved virtually complete control over their oil production facilities (Johns and Field, 1986).

In the early 1970s not all the countries of the peninsula were oil producers, with the two Yemens only finding oil in commercially exploitable quantities in the 1980s. As Table 6.4 indicates the largest producers have been Saudi Arabia, Kuwait and Abu Dhabi, and these are also the countries with the biggest proven reserves. In contrast, Bahrain, Qatar, Oman and the other emirates in the United Arab Emirates have benefitted to a much lesser extent from the inflow of oil revenues. The extent of these revenues available to the governments of the region have, though, been very large indeed, as illustrated in Table 6.5, which provides summary information concerning annual government oil revenues for selected years between 1975 and 1985. Table 6.5 also indicates the further dramatic rise in revenues following the oil price rises of 1979, when the price of Saudi Arabian Light rose from U.S.$ 12.70 a barrel in July 1978 to U.S.$ 28 in July 1980 and then to U.S.$ 34 per barrel by July 1982.

Table 6.4: Oil production in the Arabian Peninsula 1970-1985

	Oil Production million tonnes			Proved reserves at end of 1985	Ratio reserves: production 1985	Refinery capacity '000 b/d		
	1970	1975	1980	1985	million tonnes		1970	1975	1985
Abu Dhabi	33.4	67.3	64.7	41.5	4.1	98.8	–	–	–
Bahrain	3.2	3.4	2.4	–	–	–	205	250	250
Dubai	4.3	12.6	17.4	18.9	0.2	10.6	–	–	–
Kuwait	139.1	94.0	71.5	45.0	12.4	275.5	470	565	550
Oman	16.6	17.1	14.4	25.4	0.5	19.7	–	–	–
PDR Yemen	–	–	–	–	–	–	160	160	160
Qatar	17.7	21.0	22.4	14.5	0.4	17.6	–	–	–
Saudi Arabia	178.0	343.9	493.0	174.8	23.0	131.6	425	465	1115
Yemen AR	–	–	–	–	–	–	–	–	–
Neutral Zone	26.0	25.8	28.4	18.7	0.8	42.8	75	75	–

Source: British Petroleum (1980, 1986)

Table 6.5: Government oil revenues of major oil
producers in the Arabian Peninsula 1975-1985

	Oil Revenues million US $			
	1975	1978	1980	1985
Kuwait	7,500	9,500	18,300	9,000
Qatar	1,700	2,200	5,200	3,000
Saudi Arabia	27,000	38,000	104,200	28,000
UAE	6,000	8,700	19,200	12,000

Source: Middle East and North Africa (1987)

It is salient at this point briefly to put these figures into their global context, and to highlight the dilemma which they presented to the governments of the Arabian peninsula. If it had not been for the demand for hydrocarbons by the countries of north America, Japan and western Europe, which as far as oil was concerned grew from 23,580 thousand barrels a day in 1966 to 38,990 thousand barrels a day in 1973 (British Petroleum, 1986), and their inability to produce sufficient oil themselves to satisfy these requirements, then little of the change that has been seen over the last twenty years in the Arabian Peninsula would have taken place. Hydrocarbons, however, are finite in quantity, and the dilemma facing the governments of the region was therefore, having ensured a reasonable price for their exports, how best should they spend the resultant income to create economies that would continue to flourish once oil revenues had begun to dwindle. The urgency of this problem is well indicated in Table 6.4, which illustrates that even at the somewhat reduced production levels of 1985 the reserves/production ratios for some countries in the region such as Bahrain, Qatar and the emirate of Dubai are well below 30 years.

Four broad strategies have been adopted to varying degrees by the governments of the Arabian peninsula in trying to achieve sustained economic growth, and each of these has had important repercussions for agriculture and urban-rural relationships in the region.

The first, and most immediate, response to the increase in crude oil exports was the development of much greater downstream activity in the hydrocarbon sector. Having achieved control over their own production facilities in the region, the next step was one of considerable investment in refining facilities, petrochemical industries, and the final distribution networks. Kuwait, Bahrain and the People's Democratic Republic of Yemen have not dramatically increased their refining capacities, but between 1970 and 1983 Saudi Arabia's production of refined products rose from 1.8 million barrels to 111 million barrels (Kingdom of Saudi Arabia, Ministry of Planning, 1984). Kuwait has likewise paid considerable attention to downstream activities, particularly on the distribution side, with the Kuwait Petroleum Corporation (KPC) purchasing petrol stations in

Europe from Gulf Oil in 1983 and Elf in 1985, and establishing its own distribution networks. Petrochemicals are a further area of downstream diversification which has received some attention in the region. Thus Qatar, having commissioned a small refinery in 1974 established the Qatar Petrochemical Company also in 1974, and this came into production in various phases during 1980 and 1981. During the mid-1980s Saudi Arabia's new petrochemicals plants at Jeddah, Jubail and Yenbu also came into production.

Enhanced downstream activity basically increases the revenue obtainable by the governments of the region from their crude production, but it does little to diversify the economy away from oil. The other industrial developments which have been undertaken also show a heavy reliance on oil as a feedstock. Thus most countries in the region have embarked on various heavy industrial enterprises which have high energy requirements. Bahrain and Dubai, for example, have established aluminium smelters which began production in 1971 and 1979 respectively, and Qatar set up the Qatar Steel Company which began production in 1978. Likewise, hydrocarbons have also been used as feedstocks for the increasing number of fertilizer plants that have been constructed since the early 1970s, such as the Qatar Fertilizer Company which began production in 1973. Until the early 1980s most associated gas was flared off, but increasingly during the 1970s this began to be tapped as a further resource, and additional discoveries of non-associated gas, such as the massive North Field in Qatar, have provided further feedstocks which will to some extent militate against the decline of oil reserves.

A third tactic adopted by governments in the region has been investment in the capital and financial markets of the world, reinvesting income from oil exports in the expectation that returns from this investment will provide for a stable economic future. In the mid-1970s Kuwait thus established a Reserve Fund for Future Generations, to which at least 10 per cent of total revenue must be added annually by law, and it is estimated that at the end of 1979 approximately two-thirds of Kuwait's financial assets were held abroad. Bahrain has also paid considerable attention to developing a financial market, but unlike Kuwait it has concentrated on establishing its own financial centre through the creation of numerous Offshore Banking Units. Through these the government has succeeded in attracting high-level foreign investment to the island without detracting from local growth. Nevertheless as the world's financial system becomes increasingly concentrated in the three centres of London, New York and Tokyo, Bahrain's financial market is likely to find itself under increasingly difficult strains.

A final alternative has been the development of small-scale industry and agriculture, but this has generally proved to be of only limited success. Thus, as Kubursi (1984, p.71) has commented, "the relationship between the growth of oil revenues and the growth of non-oil GDP in the region has been rather tenuous". He goes on to suggest that two factors have been particularly important in limiting the growth of the non-oil sector: "heavy investments in the infrastructure of the economy that were mainly divorced from productive investments, and the sharply negative effects of oil on agriculture" (Kubursi, 1984, p.71). Much of the remainder of this chapter focuses in on the second of these two points, but four other important reasons for the lack of success of small-scale agricultural and industrial enterprises can also be suggested here. The first is that these sectors of the economy have a relatively small *absorbtive capacity* for investment. Thus agriculture was only ever going to have been able to absorb a small amount of the money requiring investment which came in to the region as a result of its oil exports. A second factor limiting the development of small-scale industry and agriculture was the *reluctance* of the indigenous inhabitants of the area to become involved in these enterprises, a point to be returned to later in this chapter. A further factor can be seen to have been the relative levels of *prestige* associated with different types of investment, with large-scale urban and industrial projects attracting much more attention than small rural schemes. Finally, it is also generally much easier to *administer* a few large and expensive schemes than it is to allocate spending over a large number of small schemes. All of these factors mitigated against real economic diversification and the development of the non-oil sector.

Despite the rhetoric of diversification it is clear that the economies of most of the countries of the Arabian peninsula are therefore still heavily oriented towards the hydrocarbons sector. But the precise balance between these four avenues of investment varies markedly between the countries, depending on their oil and gas reserves, on their endowment of other resources, and on the whims and wishes of their leaders. Saudi Arabia, Qatar and the United Arab Emirates have generally followed the path of heavy industrialisation, Kuwait and Bahrain have emphasised the financial investment element, with Bahrain also opting for some industrialisation, whereas the Yemens and Oman have until recently lagged behind considerably in terms of economic change. The constant feature of all these patterns of development, though, is that they have been primarily industrial in nature and have been associated with forces giving rise to rapid urbanisation. The consequences of these changes for the distribution of population and social structure have been dramatic and have played a major role in re-orienting the pattern of urban-rural interaction, to which attention now turns.

Social and demographic change: the urban-rural balance

The major constraints on agricultural development in the Arabian peninsula have already been shown to be environmental. In contrast, the main factors inhibiting industrialisation have been demographic, in the form of the absolute lack of population and also of their lack of suitable skills. Much of the explanation for the nature of recent economic change in the region can thus be attributed to the ease with which labour could be imported in comparison to the expense and difficulty of creating a suitable environment for agriculture.

Table 6.1 highlighted the very low populations of all of the countries of the Arabian peninsula apart from Saudi Arabia and the Yemen Arab Republic at the beginning of the 1970s. This can be seen to have limited industrialisation in two important ways. It did not provide a sufficient market for the products of the newly proposed industries, nor did it supply a sufficient labour force. The marketing problem has remained a significant constraint on industrial development, and overcapacity in the region is proving to be particularly troublesome as further industrial projects come on stream.

The labour migration that enabled this industrial capacity to be achieved is nevertheless also crucial for an understanding of the changed urban-rural relations to be found in the region. Identifying the precise extent of labour migration, much of which may not be legal, is far from easy, but in a major study of migration in the Middle East in 1975, Birks and Sinclair (1980) illustrated the broad pattern of international migration in the region. Table 6.6, which is derived from this study, illustrates three important features: firstly, the sheer magnitude of the amount of migration totalling some 1,389,052 workers in 1975; secondly, the flow of workers from the poor countries of the region without extensive oil reserves, such as the Yemens and to a lesser extent Oman, to the oil-rich countries of Saudi Arabia and the UAE; and thirdly, the large number of Asian immigrants, particularly to the UAE. This pattern of migration continued through the remainder of the 1970s, and by 1980 it can be estimated that in Qatar the indigenous population represented only about 20 per cent of the total, in the UAE about 25 per cent of the total, and in Kuwait about 40 per cent of the total. Bahrain and Saudi Arabia on the other hand retained much higher levels of indigenous population with only about 30 per cent of their populations being immigrants. Apart from the social influence of this extensive immigration, it had two other important repurcussions for a study of urban-rural relations. The first of these is that the vast majority of the immigrants were employed in industry, either in the oil-related sector or in construction. The significant feature of this is that most of these industries were located in urban areas, and as a result urban food demand increased dramatically. Secondly, though, the remittances brought back by migrant workers to Oman and the Yemens, provided the opportunity for some form of investment in the rural areas of these countries.

At the same time as this labour immigration was taking place two other processes were operating to give rise to a more highly urbanised population. On the one hand the governments, particularly in Saudi Arabia, Oman and the United Arab Emirates, were embarking on schemes to resettle the nomadic elements of their populations in small towns or villages, and on the other hand as the towns began

Table 6.6: Migrant workers in the Arabian Peninsula, 1975

TO	FROM Egypt	Iraq	Palestine Jordan	Lebanon	Yemen PDR	Oman	Sudan	Syria	Yemen AR	Pakistan	India	Total Migrant Workers
Bahrain	1,237	126	614	129	1,122	1,383	400	68	1,121	6,680	8,943	29,285
Kuwait	37,558	17,999	47,653	7,232	8,658	3,660	500	16,547	2,757	11,038	21,475	208,001
Oman	4,600	-	1,600	1,100	100	-	500	400	100	32,500	26,000	70,700
Qatar	2,850	-	6,000	500	1,250	1,870	400	750	1,250	16,000	16,000	53,716
Saudi Arabia	95,000	2,000	175,000	20,000	55,000	17,500	35,000	15,000	280,400	15,000	15,000	773,400
UAE	12,500	500	14,500	1,500	4,500	14,000	1,500	4,500	4,500	100,000	61,500	251,500
Yemen AR	2,000	-	200	-	-	-	-	150	-	-	-	2,450

Source: Birks, J.S. and Sinclair, C.A. (1980)

to be developed these provided more opportunities for the traditional village population to find forms of employment with better remuneration than agriculture. The history of attempted settlements of nomadic populations in Arabia goes back at least as far as Ibn Saud's al-Hijar settlement scheme (Joffe, 1985), but the major influence on the collapse of the nomadic lifestyle in Saudi Arabia was the combination of increased opportunities of urban employment with the issuing in 1968 of the Public Lands Distribution Ordinance which enabled collective land to be turned into private property. As Joffe (1982, p.221) has commented "For the *bedu*, the only solutions were to join in the new private ownership schemes, or to leave the sector altogether". Janzen (1983, 1986) has likewise surveyed the effect of the Omani government's policy of resettling the nomadic population of Dhofar, through the use of revenues derived from its burgeoning oil industry. However he concludes pessimistically that "In view of the unhealthy employment policy which has taken shape in Dhofar, it seems doubtful whether the nomadic population, which has become accustomed to earning easy money through the state-initiated assistance measures, is in any way willing to play an active part in the development of new branches of the economy and the modernisation of the traditional ones. It is to be feared that no productive work can be expected from the nomads in their traditional forms of activity, beyond whatever may be required for purposes of subsistence" (Janzen, 1986, p.287).

Having surveyed the broad patterns of recent industrial change, and its effects on population levels and labour movements it is now possible to look in detail at the resultant changes in agricultural production and changes in urban-rural food flows.

The Urban-Rural Context of Agrarian Change

Broadly speaking it is possible to see the changing nature of urban-rural relations in the Arabian peninsula over the last twenty years as having followed a three phase chronology. In the late 1960s and early 1970s, as the oil industry grew, there was an influx of foreign immigrants to the new cities of the region. Domestic agriculture was unable to support this increased demand, and food imports on a massive scale were required. In the second phase in the mid 1970s, agriculture can be seen as having become increasingly less attractive and viable. Food imports were undercutting the prices of domestic produce, and employment in the new towns and cities was seen as being more desirable than the traditional life of farming. Moreover, increased extraction of underground water resources to satisfy the growing industrial and urban domestic demands was leading to the lowering of water tables and increased salinity of the aquifers. However, from the late 1970s and early 1980s the balance had swung back to some extent towards the rural economy, and considerable investments have been made in the agricultural sector, under the rhetoric of a drive to self-sufficiency. The precise processes involved in each of these phases can now be elucidated in more detail.

Table 6.1 has sketched the broad outline of population growth in the region since 1970 and its increasingly urban nature. In trying to relate these figures to changes in agricultural production and food imports the data limitations can not be overstressed, but Tables 6.7 and 6.8 provide a very approximate indication of changes in the agrarian economy over this period. Despite the unreliability of the data, two clear impressions can be gleaned from Table 6.7 concerning the nature of the food imports necessary to sustain the growth of urban population. The first is that food imports increased dramatically to provide sustenance for the largely immigrant labour force, and the second is that, as the economies of the region have become more industrialised over the same period, the value of food imports as a percentage of total imports has diminished. This paradoxical situation has given rise to an interesting dilemma for those in planning departments in the region, because it would seem that, as the countries' economies are becoming more urban and industrialised they are in turn becoming more dependent on manufactured imports. Yet at the same time there is an increasing need to reduce the total amount of food imports. Since the economic downturn of the early 1980s and the collapse of oil prices in 1986, it is also pertinent to note that population levels have ceased to grow so rapidly, and therefore that the dramatically increased pressure on food resources that was felt in the 1970s has also diminished.

Table 6.7: Food and live animal imports at current prices

| | 1970 | | 1975 | | 1982 | |
	Value '000 US$	Percentage total imports	Value '000 US$	Percentage total imports	Value '000 US$	Percentage total imports
Bahrain	28,266	11.3	58,098	5.6	180,883	5.5
Kuwait	105,851	17.0	365,322	15.3	1,018,163	12.2
Oman	–	–	77,679	–	346,929*	13.8*
Qatar	–	21.8	51,130	13.0	168,641	9.2
Saudi Arabia	180,257	26.8	544,812	13.7	4,668,704	11.2
UAE	–	–	257,680	9.4	711,715	8.7
Yemen AR	–	–	126,338	41.9	–	–

Source: United Nations (1970-1984)

Note: * figures refer to 1984

Table 6.8: Food and livestock production

	Arable and Permanent Crops '000 hectares				Total cereal Production '000 tonnes			Poultry '000 head			Total Cattle '000 head	
	1970	1975	1980	1984	1975	1980	1985	1970	1980	1985	1970	1985
Bahrain	2	2	2	2	-	-	-	110	775	1000	4	6
Kuwait	1	1	1	2	-	-	-	4475	5941	8000	5	160
Oman	36	307	41	43	5	3	2	473	462	1000	68	130
PDR Yemen	130	142	156	167	195	207	115	1237	1520	2000	80	96
Qatar	1	2	2	3	-	1	1	60	350		5	7
Saudi Arabia	859	1105	1105	1156	570	285	1811	3000	5500	34000	225	540
UAE	11	12	13	15	-	1	4	85	2000	4000	16	31
Yemen AR	1300	1347	1351	1351	1975	793	580	2801	3400	15000	842	950

Source: Food and Agriculture Organization (1979-1985)

Nevertheless, in summarising the nature of food flows since 1970 it is evident that they do not reflect a major shift in the traditional pattern of exchange. As was pointed out earlier in this chapter, the towns of the Arabian peninsula have *always* been importers of food from overseas and the recent industrialisation and rapid urbanisation have merely accentuated an already existing pattern rather than introduced a new one.

Table 6.8 indicates another feature of agrarian change during the early 1970s, and this is that domestic agriculture does not appear to have responded at all rapidly to the increases in urban demand for food. While areas of land under cultivation and levels of production did increase somewhat, there was no massive expansion of agriculture until the late 1970s. This can be seen as a result of at least two factors. On the one hand there was a general shift of population away from the rural areas, characterised by hard toil in an inhospitable environment, to the apparently easier and more remunerative employment in the towns. Put at its simplest, the environmental constraints made it extremely difficult *rapidly* to increase agricultural production. As Kubursi (1980, p.72) has noted, the "inflow of subsidised imports depressed food prices domestically to the extent that domestic farmers' opportunity costs became substantially high and many of them left the farms for the cities. The decrease in food prices was matched by very high costs of production and labour, so much so that farmers were caught in a cleft stick - low prices for their products and high costs for their inputs". A second factor behind the slow development of agriculture was the lack of attention paid to it in official or unofficial development plans. Kubursi (1980) has thus noted that throughout the period 1970-1980 the governments of the Gulf Co-operation Council (GCC) countries all spent under 5 per cent of their total development expenditure on agriculture. The drive to rapid industrialisation and urbanisation left agriculture in a quiet backwater, to survive as best it could.

Bowen-Jones and Dutton (1983, p.7) have perfectly captured the situation prevailing in most of the region, apart from in the Yemens, when they stated that "everywhere else there has appeared the paradox that whilst consumer demand for agricultural products has soared, qualitatively and quantitatively, the survival imperative for domestic farm production has almost vanished". Agricultural experiment-ation stations, such as that at Digdaga in Ras al-Khaimah, have been constructed in many parts of the region, farmers are heavily subsidised and encouraged to expand agricultural production, new marketing schemes to deliver the food to urban areas have been developed as in Abu Dhabi (Unwin, 1981, 1983b), but until the late 1970s these had relatively little overall influence on the expansion of production. Instead the two traditional systems of agricultural production, the one settled and the other nomadic, have gradually declined at the expense of highly capitalised, large-scale farming enterprises. It has thus only been through the injection of large amounts of capital in the form of new deep wells, irrigation systems and fertilizers that the fundamental environmental constraints on agriculture have been *temporarily* overcome. By and large, though, it has not been the traditional farming and pastoral communities that have benefitted directly from these changes.

There have also been important differences between the fortunes of the traditional settled crop producing communites and the nomadic livestock pastoralists. The most dramatic changes have taken place in livestock production, and the effects on the way of life of the nomadic bedouin, commented upon by Janzen (1986) for Dhofar, have been replicated throughout the peninsula. Livestock production has shifted from being extensive and rural-based to being highly intensive and urban in location, and as a result vast tracts of traditional pasture land are now no longer used for any productive purpose. As Bowen-Jones and Dutton (1986, p.8) have commented, "The most extreme development of this kind is in poultry and egg production. The availability of capital, the absence of import controls, the buoyant free market demand associated with high incomes and dietary preferences, all these have created conditions appropriate for a commercial factory style production unit approach" (Table 6.8). To this can be added the observation that as the feed for the poultry is almost totally imported, and as much of the population lives in these urban port agglomerations, these new poultry units are primarily urban in location. Schemes for cattle and milk production have also flourished, with one of the earliest being the dairy unit at Digdaga in

Ras al-Khaimah. This was begun by the British in 1957 with a small herd of Friesian cows, but by the early 1980s its activities had expanded considerably with 600 head of cattle providing between 12,000 and 15,000 litres of milk a day. Unlike with poultry production, many of the dairy schemes have been associated with their own fodder production enterprises. Thus the Digdaga dairy farm in the early 1980s was producing fodder from a 250 hectare site at Hamraniyah irrigated by a system of sprinklers every 15 metres. The resultant influence on the environment of such intensive irrigation was that in the first year the water table fell by about one metre, and in subsequent years it has continued to do so.

Turning to crop production, the traditional way of life has to some extent continued, particularly in the areas such as Jizan and the mountains of Oman. But in this sector as well, intensive, highly capitalised agricultural schemes have been developed since the end of the 1970s. At least four factors associated with the increased urbanisation that has been seen in the region since 1970 have militated against the success of small-scale irrigated or well-fed agriculture in the region. Firstly the urban employment prospects for farm owners acted as an incentive for them to leave productive farming. Nevertheless, another side of this coin is that many of these landowners have retained their 'farms' as small 'date gardens' to which they go to relax at week-ends or in the evenings. Thus, as long as they are retained by their owners primarily for purposes of leisure there is unlikely to be any major change towards productive agricultural enterprise on these 'farms' in the near future. A second factor concerns the nature of the agricultural labour force. As indigenous labour moved to the towns in search of better prospects in the 1970s, most of the agricultural labour force came to consist of relatively unskilled immigrants, who had little motivation for high levels of productivity. To exacerbate matters further, the high water demands of the growing cities and industrial enterprises during the 1970s led to a dramatic lowering of water tables and increasing soil salinity. This was particularly the case in coastal areas where sea water was able to encroach into the previously freshwater aquifers as in Ras al-Khaimah and Bahrain, where large expanses of date gardens died as a consequence. A fourth factor is that the wealth generated by oil enabled large amounts of food to be imported relatively cheaply to feed the urban population, and this in turn undermined the viability of domestic crop production. A complex series of relationships therefore existed between agricultural production and the forces leading to an increasingly urban population, and the net effect was to make small-scale crop production increasingly unviable.

The alternative, as with livestock production, was to use some of the wealth generated from oil to finance capital intensive large-scale agricultural enterprises. Two particular influences can be seen as having played an important role in giving rise to this change. The first is that with the foundation of the Gulf Co-operation Council in 1981 came an increased awareness of the need for regional co-operation in all aspects of economic development, and in particular of the fragility of the economies of member countries as they then existed. It is from this date that increasing mentions of the drive to agricultural self-sufficiency are to be found in the rhetoric of government ministers. At about the same time there was also a growing awareness that if the pumping of oil had transformed the industrial side of their economies, why should the pumping of water not transform their agriculture? While vast strides have been made since the late 1970s in the provision of desalination plants in the region, much of the water for agriculture remains drawn from fossil acquifers. Although most authorities (Bowen-Jones and Dutton, 1983) accept that in Saudi Arabia these aquifers will last for several hundred years, their depletion remains of the utmost concern, and in some of the peripheral countries of the peninsula such as Bahrain, Qatar and the United Arab Emirates the depletion of aquifers has already had a serious influence on the agrarian economy. The cost of water is heavily subsidised in the region, and at real costs it would make little sense to use desalinated water for agriculture. To make matters worse, the inappropriate and uncontrolled application of water has meant that in many areas soils have become waterlogged and increasingly saline. Nevertheless as Joffe (1986) has noted for Saudi Arabia, the drive for this kind of agricultural 'development' continues, and it is such large-scale schemes that have been responsible for the dramatic increases in wheat production in Saudi Arabia noted in the quotations at the beginning of this chapter. Flows of food from rural areas to the towns are in absolute terms greater today than they have ever been, but for how long they will last remains to be seen.

Conclusion

In conclusion it is possible to make three general assertions concerning the changes that have taken place in terms of urban-rural relations associated with agriculture in the Arabian peninsula over the last thirty years. The first is that prior to the discovery of oil there was a considerable dislocation between the urban and rural economies in much of the region. Only in the south-west of the peninsula can there to any extent be said to have been an urban economy sustained by the marketing of locally grown agricultural products. Elsewhere the majority of towns survived essentially on the importation of food from overseas. The rapid urbanisation that followed in the wake of the exploitation of the region's oil reserves can thus be seen as having reinforced and exaggerated traditional patterns of food imports, rather than having destroyed imagined urban-rural linkages.

Nevertheless, the second conclusion is that by offering new forms of employment, and new sources of income, industrialisation based upon oil and the way of urban life that became associated with it, did contribute greatly to the destruction of traditional rural lifestyles. This has been particularly marked with the virtual disappearance of pastoral nomadism, and the consequent cessation of productive use of vast tracts of desert. Traditional forms of settled agriculture as described in the pages of Lorimer's early 20th century account of the region have also largely been superseded. In their place has emerged, but only recently, a new capital intensive and environmentally high-risk form of livestock and cereal production. The availability of the requisite inputs, including massive amounts of water from fossil aquifers, has enabled the output of grain from parts of Saudi Arabia even to exceed the domestic requirements of that kingdom. Not only have rural areas therefore now begun to provide food for the towns of the region, and thus to generate a whole new series of rural to urban flows, but also it is possible to envisage, at least in the short term, exports of some agricultural products. The third conclusion to note, though, is that there have been two casualties as a result of this transformation. On the one hand the traditional farming population, apart from in Oman and parts of the Yemens, has all but disappeared, and with it so have the farming skills accumulated over centuries. On the other hand, and perhaps more worryingly, many of the soils, once farmed continually for many years, have now become so saline that any attempts at reclaiming them are likely to be prohibitively expensive. In the advance of 'progress' the environment has been the loser.

REFERENCES

Allan, J.A. (1981) 'Renewable natural resources', in: Clarke, J.I. and Bowen-Jones, H. (eds.) *Change and Development in the Middle East: essays in honour of W.B. Fisher*, Methuen, London

Beaumont, P., Blake, G.H. and Wagstaff, J.M. (1976) *The Middle East: a geographical study*, Wiley, London

Birks, J.S. and C.A. Sinclair (1980) *Arab Manpower: the Crisis of Development*, Croom Helm, London

Bonine, M.E. (1980) 'The urbanization of the Persian Gulf nations', in: Cottrell, A.J. et al. (eds.) *The Persian Gulf States: a General Survey*, The Johns Hopkins University Press, Baltimore, 225-78

Bowen, R. le L.B. (1951) 'The pearl fisheries of the Persian Gulf', *The Middle East Journal* 5, 161-80

Bowen-Jones, H. and Dutton, R. (1983) *Agriculture in the Arabian Peninsula*, Economist Intelligence Unit, London

British Petroleum (1980) *BP Statistical Review of World Energy 1980*, British Petroleum, London

British Petroleum (1986) *BP Statistical Review of World Energy 1986*, British Petroleum, London

Clarke, J.I. (1981) 'Contemporary urban growth in the Middle East', in: Clarke, J.I. and Bowen-Jones, H. (eds.) *Change and Development in the Middle East: essays in honour of W.B. Fisher*, Methuen, London

Drysdale, A. and Blake, G.H. (1985) *The Middle East and North Africa: a political geography*, Oxford University Press, New York

Dutton, R. (1985) 'Agricultural policy and development: Oman, Bahrain, Qatar and the United Arab Emirates', in: Beaumont, P. and McLachlan, K. (eds.) *Agricultural Development in the Middle East*, John Wiley, Chichester, 227-40

El Mallakh, R. (1968) *Economic Development and Regional Cooperation: Kuwait*, University of Chicago Press, Chicago

El Mallakh, R. (1981) *The Economic Development of the United Arab Emirates*, Croom Helm, London

Food and Agriculture Organization (1971-1985) *F.A.O. Production Yearbooks, Vol.25-Vol.39*, Food and Agriculture Organization, Rome

Janzen, J. (1983) 'The modern development of nomadic living space in southeast Arabia - the case of Dhofar (Sultanate of Oman)', *Geoforum* 14, 289-310

Janzen, J. (1986) *Nomads in the Sultanate of Oman: Tradition and Development in Dhofar*, Westview, Boulder

Joffe, E.G.H. (1985) 'Agricultural development in Saudi Arabia: the problematic path to self-sufficiency', in: Beaumont, P. and McLachlan, K. (eds.) *Agricultural Development in the Middle East*, John Wiley, Chichester, 209-26

Johns, R. and Field, M. (1986) 'Oil in the Middle East and North Africa', *The Middle East and North Africa 1987*, Europa Publications, London

Kingdom of Saudi Arabia, Ministry of Agriculture and Water (1986) *Agricultural Development in the Kingdom of Saudi Arabia*, Ministry of Agriculture and Water, Riyadh

Kingdom of Saudi Arabia, Ministry of Planning (1984) *Achievements of the Development Plans 1390-1404 (1970-1984)*, Ministry of Planning, Riyadh

Kubursi, A.A. (1984) *Oil, Industrialization and Development in the Arab Gulf States*, Croom Helm, London

Lorimer, J.G. (1970) *Gazetteer of the Persian Gulf, Oman and Central Arabia*, Government Printing House, Calcutta, first published 1908-1915; reprinted Gregg, Farnborough

McLachlan, K. (1980) 'Oil in the Persian Gulf area', in: Cottrell, A.J. et al. (eds.) *The Persian Gulf States: a general survey*, The Johns Hopkins University Press, Baltimore, 195-224

Middle East and North Africa (1987), Europa Publications, London

Mountjoy, A.B. (1982) *Industrialisation and Developing Countries*, Hutchinson, London, 5th edition

Nugent, J.B. and, Thomas, T.H. (eds.) (1985) *Bahrain and the Gulf*, Croom Helm, London

Pridham, B.R. (ed.) (1985) *The Arab Gulf and the West*, Croom Helm, London
Raswan, C.R. (1930) 'Tribal areas and migration lines of the north Arabian bedouins', *American Geographical Review* 20, 494-502
Riyadh Municipality (1986) *Riyadh ... A contemporary city of cultural heritage*, Riyadh Municipality, Riyadh
Sweet, L.E. (1964) 'Pirates or polities? Arab societies of the Persian or Arabian Gulf, 18th Century', *Ethnohistory* 22, 262-80
United Nations (1970-1984) *Yearbooks of International Trade Statistics*, United Nations, New York
Unwin, P.T.H. (1981) *Agriculture in the United Arab Emirates: a preliminary report*, Department of Geography, University of Durham, Durham
Unwin, P.T.H. (1982) 'The contemporary city in the United Arab Emirates', in: Serageldin, I. and El-Sadek, S. (eds.) *The Arab City: its character and Islamic cultural heritage*, Arab Urban Development Institute, Riyadh, 120-31
Unwin, P.T.H. (1983a) *The Arab Contribution to Agriculture in the north-eastern Arabian Pensinula*, The Third International Symposium on the History of Arab Science, Kuwait
Unwin, P.T.H. (1983b) 'Agriculture and water resources in the United Arab Emirates', *Arab Gulf Journal* 3, 75-86
Wilkinson, J.C. (1977) *Water and Tribal Settlement in south-east Arabia: a Study of the Aflaj of Oman*, Clarendon Press, Oxford
Wilson, R. (1977) *Trade and Investment in the Middle East*, Macmillan, London.
Wrigley, E.A. and P.S. Schofield (1981) *The Population History of England 1541-1871: a reconstruction*, Edward Arnold, London
Yapp, M. (1980) 'The nineteenth and twentieth centuries', in: Cottrell, A.J. *et al.* (eds.) *The Persian Gulf States: a general survey*, The Johns Hopkins University Press, Baltimore, 41-69
Zahlan, R.S. (1979) *The Creation of Qatar*, Croom Helm, London

7

COMMERCIALISATION, DISTRIBUTION AND CONSUMPTION: RURAL-URBAN GRAIN AND RESOURCE TRANSFERS IN PEASANT SOCIETY

BARBARA HARRISS

The nature of commercialisation and the emergence of marketing systems have crucial significance for economic and social change. Moreover, the processes involved in exchange relations, and in the extraction and use of surpluses, have important social repercussions which are reflected in certain circumstances in net flows of surplus from rural to urban areas. This chapter develops these ideas first in a theoretical framework and then through an analysis of detailed case studies at the village level in Hausaland (semi-arid West Africa) and at the regional scale in Tamil Nadu (semi-arid South India). In conclusion it explores the relationships between commercialisation and nutrition levels in peasant households.

Commercialisation and market exchange of grain in the rural-urban context

The production of grain is essential for the reproduction of almost all societies, and the commercialisation of grain is essential to the capitalist commoditisation of labour, which historically it has antedated. The process of commercialisation has typically begun with output markets; but the existence of markets and of sales to, and purchases from, them is no precise indicator of relations of production or of levels of development (Chaudhuri, 1979; Friedmann, 1983; Bharadwaj, 1985). While we have no quarrel with the notion of the primacy of self-provisioning as the objective of peasant households, for many years, and over a millennium in some regions of India, peasants have had to buy. In so doing they have regularly had to buy outside their villages, and they have engaged in direct and indirect transactions with an urban world.

The contemporary commercialisation of grain is part of a wider process of appropriation of surplus in the form of products. To locate this contemporary process it is necessary first to outline the ways in which control over resources is transferred out of the hands of people directly engaged in their production. One way that this transfer is achieved is through *rent* on land, water or property. The development of land mortgage implies an increase in surplus appropriation and an increase in the commercialisation of commodities for mortgage repayment which is actually the payment of disguised rent. Another mechanism of transfer is through *usury*: the repayment, in cash, kind or labour, of money or kind loaned at high interest rates not reflective of a competitive money market. These are characteristic precapitalist forms of surplus appropriation, where rent is not based on land values or productivity, and where extra-economic coercion is used to set levels of interest.

Another means of surplus appropriation is *taxation* on land, property, income, people, or on goods sold (Rao, 1985). Yet another is via the exploitation of *labour* power where the wage, which may be sufficient for nutritional subsistence and sometimes also for the reproduction of the labourer's family, is less than the value of the production of that labour. The exploitation of labour can be increased under conditions where wages are reduced because the labourer has land of *his* own and can provide *his* own means of subsistence, through his unvalued domestic productive labour, or, commonly through that of adult women in the household (Deere, 1979; Young *et al.*, 1981).

Commercialisation is manifested in increased *market exchange* for three main purposes:

(i) In order to purchase *goods not produced by the household*. Grain may be sold for cattle, salt, mill cloth, kerosene and jewellery. Grain may be purchased when grain production has failed.

(ii) In order to purchase goods *produced but not owned by the household*. Grain will be purchased by landless labourers paid in cash for their work in the production of grain. Their labour power is commercialised. Grain will be purchase or bartered in rural areas by labour paid in cash *or in kind* for their work in the production of non staples, such as fruit, vegetables and spices, and agro-industrial crops, such as silk, cotton and rubber. These first two purposes may also be effected by exchange in kind, or by barter, though commercialistion is characterised by increasing use of money in exchange.

(iii) In order to *obtain cash.* Grain may be sold for cash in order to pay taxes, in order to pay rent, and in order to repay debts, for ceremonies and for consumption.

Likewise, surplus is extracted through market sales under three broad conditions:

(i) Where the agricultural marketing system is itself reasonably competitive and efficient, but where the power of non-agricultural classes is sufficient to manipulate price levels in such a way that the long term barter terms of trade are turned against the agricultural sector.

(ii) Where monopolistic merchants are able to depress their purchase prices (peasant producers' selling prices) below what they would be in a competitive market.

(iii) Where State pricing policies and other State interventions practise prices below those of private markets, and compel peasants to sell to the State.

Merchants' capital

Classically the role of merchants, who are frequently urban-based, in the process of commercialisation has been recognised to be profoundly contradictory. On the one hand, since there are no activities connected with buying and selling which change the nature of commodities and increase their use value, the usefulness of merchants consists in the fact that "a smaller part of society's labour power and labour time is tied up in this unproductive function" (Marx, 1974, vol.1, p.134). So, firstly trade is considered to be *unproductive but necessary* because "the process of (social) reproduction itself includes unproductive functions" (Marx, 1974, vol.1, p.134). Secondly trade has a *retrogressive* role because of its historical tendency to depend on precapitalist social relations of production for the exploitation of labour and the extraction of surplus, and because of its tendency to develop and reinforce dominating exchange relationships which may verge on monopolies. Relatively depressed prices to producers, characteristic of mercantile monopolies, thwart the development of production upon which the increase in commodities traded ultimately depends. On the other hand the commercial firm is usually more than the embodiment of merchants' capital. Few traders restrict their activities to buying and selling. They engage in transport and storage and may process goods as well. Transport is a productive activity: "the use value of things is materialised only in their consumption and this may necessitate a change of location of those things; hence may require an additional process of production in transport" (Marx, 1974, vol.2, p.153). Processing likewise represents "an interruption of the process of circulation for productive purposes" (Marx, 1974, vol.2, p.104). As regards storage, Marx (1974, vol.2, p.104) argued, "storage enhances prices without contributing to use value". In locking up resources to prevent loss, storage is unproductive, but it is nonetheless necessary to the organisation of the supply of goods. But we can borrow his argument that resources used in the maintenance of machinery are productive, and apply it to storage to the extent that resources are used to stop commodities from deteriorating. So, mercantile firms undertake permutations and combinations of productive activities as well as unproductive but necessary ones and (as we shall see) unproductive and unnecessary ones. And further, trade has a *progressive* role because it enables an increase in commodity production, the monetisation of economies, and the integration of regions. It is through the mechanism of prices, and the way in which price signals representing demand are transmitted through the marketing system that producers respond with varying degrees of sensitivity by allocating resources in market-oriented production. The functioning of competitive marketing systems therefore can lead to efficient resource allocation within agriculture, and to interregional resource allocations in response to the principle of comparative advantage (Bressler and King, 1970; von Oppen, 1985), according to which aggregate agricultural production will be maximised. In most instances the marketing system is articulated around a few major urban centres which provide the concentrated foci of activity of these mercantile firms.

The mercantile sector, urban agent in the process of commercialisation, may function to promote unproductive accumulation, to preserve precapitalist production relations, or the reverse: to promote productive accumulation and break the fetters of precapitalist relations (Bharadwaj, 1985). It may also embody forces working simultaneously in both directions. There is nothing inevitable about commercialisation, and many historical instances of atrophy, decline or reversion to subsistence can be found (see Pouchepadass, 1981, for Bihar; Baker, 1984, for Madras; Mishra, 1981, for Bombay). It is perfectly possible for the commercialisation of foodgrains to take place without a commensurate expansion in production, let alone the more common case where production expands but consumption by direct producers falls. There is no necessary connection between commercialisation and agricultural growth. Pre-existing production relations are moulders of the structure and behaviour of commodity markets. Urban-based commercialisation need not necessarily by itself generate growth in rural production.

Exchange relations

The transaction between rural producer and urban trader constitutes a social relationship, a relation of exchange. Within peasant society, exchange relations can be immensely varied. Bharadwaj (1974, 1985) has given us an archetypical set. Large cultivators dominate contributions to the marketed surplus and make free decisions about participation in the market, decisions which in turn set the terms and conditions for exchange. Middle sized cultivators are responsive to the prices created by the transactions of large cultivators. Small peasants cover subsistence objectives, selling only in years when there is a physical surplus over their needs; while marginal peasants produce insufficient grain to provision their households, compensating for this deficit by selling their labour and buying grain with their wages. By virtue of the fact that this latter group of peasants is deeply involved in commercial transactions and dependent as consumers upon markets for grain, they are sometimes regarded as disguised proletarians, but they differ from 'full time' labourers because of their land and thus their vulnerability as producers to prices established by the relations between large cultivators and merchants (Adnam, 1985).

Whether or not the small peasant enters the market only when there is a residual surplus over his subsistence needs is debated. Many writers have identified a dichotomy between normal, market-incentive commercialisation resulting from the increased production of goods surplus to domestic consumption needs on the one hand, and 'forced', 'superficial' or distress commercialisation on the other (see, for example, Deere and de Janvry, 1979; Guillet, 1981; Nadkarni, 1979; Harriss *et al.*, 1984). Distress sales are made by peasants who have no option either over the timing, or the intermediary to whom they sell, or the quantity sold, or all three. Typically part of the consumption stock is sold prior to harvest or during the post harvest glut at the lowest prices to meet urgent cash needs. Distress purchases or buy-back for consumption at the highest pre-harvest prices complement distress sales. The terms of trade faced by households compulsively involved in superficial commercialisation will be disadvantageous relative to those faced by those with market-incentive commercial relations (Gore, 1978). It may be hypothesised further, as Bhaduri and Sud (1981) have done, that the mix between normal and forced commerce may reflect a mix of two surplus extracting classes: "the rich agriculturalists and 'progressive' landlords who contributed to the normal process of commercialisation by augmenting their production and the trader-moneylender class who contributed to the 'forced' process of commercialisation by forcing peasants to depend more heavily on the market, mainly through the mechanism of debt" (Bhaduri and Sud, 1981, p.5).

The use of surplus

Once extracted, resources are then put to use either directly or via channels of savings and investment in private or state sector infrastructure, industry and commerce. By and large these are urban based in nature, and this therefore engenders a transfer of surplus away from the rural areas where it is generated. They may also, though, be returned to agriculture, used for urban or rural welfare or in what is known as conspicuous consumption, for which productive activity of some part is almost always necessary, or in support of 'urban' sections of society such as the bureaucracy and the military (see Byres, 1972, 1974; Nadkarni, 1979). The process is bafflingly diverse. The role of a marketing system as a conduit for resources to flow between agriculture and other sectors, and from rural to urban areas, can facilitate a net extraction of resources from peasant producers, and can in turn inhibit the production of grain.

The interlocking of markets

The trading of grain has to be financed somehow, if it is carried out by means other than by barter. So does the production of grain, if it involves purchased inputs. Markets for grain and money can be inter-locked not simply by their flows but also by the structure of their control. In an interlocked market the dominant party links terms of contract in more than one type of market, and thereby enhances his power to appropriate surplus. The weaker party loses his freedom to choose in the markets which have been linked for him in this way. Recent empirical evidence for northern India casts doubt on the prevalence of links between land and labour, and land and credit markets (Bardhan and Rudra, 1978). But it is quite commonly found that the interlocking of money and commodity markets engenders chronic indebtedness and results in exploitation of producers by landlord traders and moneylender-traders through mechanisms such as high and disguised interest charges on kind payments for loans (Harriss *et al.*, 1984; Rao and Subbarao, 1976; Sarkar, 1979, 1981). Others have contended that high interest rates in the unorganised money markets to which food producers have to take recourse during their production season have components of monopoly profit (Nisbet, 1967; Lipton, 1977; Michie, 1978), and vary according to exchange relations between peasants and traders, with small peasants receiving the lowest prices and being charged the highest interest (Watson and Harriss, 1984;

Prasad *et al.*, 1986). High interest rates have been justified because of a combination of inelastic and rising demand for money on the one hand, and the costs of borrowing, of administration and of risks of default on the other (Bottomley, 1964; Ghatak, 1975). Bhaduri (1977), however, has cautioned that since relations between participants in the money markets are highly personalised, and since securities for loans of money are usually valued in an arbitrary way then interest rates can be set deliberately to influence the repayment behaviour of the borrower. It would be foolhardy to wish away these different explanations of the structure and functioning of the money market as the result of ideology. It is obvious that there is much diversity to observe and explain.

The repayment of loans in kind, by labouring on the lender's land or by returning money with commodities like food at a time specified by the lender, such as immediately post harvest when 'market' prices are in any case low, illustrate both the way interest rates can be deliberately set, and the crucial interlocking of money and commodity markets. Such interlocked exchange relations may force peasants compulsively involved in the market to hold onto their land, so that neither land nor labour market can fully form. Furthermore, because of such interlocking, it can be quite misleading to calculate the efficiency or profitability of commodity marketing and moneylending in isolation from each other (see Harriss, 1980, for a critique). Taken together, the way money is marketed among traders and between them and producers can increase the pace at which resources are removed from agricultural - especially food - production.

Socio-spatial interlocking

Social relationships of exchange will reveal themselves geographically in patterns of market places and a tracery of commodity flows, both between different rural areas and also between urban and rural areas. But some patterns are hard to see, and it is wrong to infer social process from spatial pattern. Petty commodity production may be associated with decentralised unspecialised trade carried out by farmer-traders, often women, not always using cash, and sited within the house or compound. Small scale intra-village distribution may also occur (see Hill, 1969, 1971, for West African examples). But the house or field may be the site for exploitive pre-harvest contracts between peasants and specialist wholesale merchants linking grain and money markets through consumption loans (see Harriss, 1984, for South Indian examples). These wholesale merchants are generally urban based, and it is therefore by means of these transactions that it is possible to see the transfer of commodity and financial surplus away from those in rural agricultural areas.

Commonly in peasant societies rural markets are periodic and in a given region their spatio-temporal structure is complex, different subsets of marketplaces being activated each day. A periodic market may contain a variety of exchange relationships or may be specialised and specific. Local, unspecialised redistribution via competitive petty trading where prices are formed by haggling is found from Middle America (Nash, 1967) through Senegal (Minvieille, 1976) to central India (Wanmali, 1981). More widespread distribution may be effected by itinerant or mobile traders (for bibliographies on which, see Bromley, 1974, 1979). The extractive bulking of commodities from periodic markets through a local wholesale town may resemble a solar system (for a famous example of which in western Guatemala see Tax, 1953). A further alternative is for rural periodic markets to form the lowest, multi commodity ring of a hierarchy of retail centres or central places which are otherwise essentially urban in nature (Gormsen, 1976; Smith, 1976; Smith, 1978). Not only are periodic markets the geographical signposts to a variety of exchange relations, but they are also the venue for 'social activities' by which is meant entertainment, politicking, the exchange of news, disputes and agreements between kin, the meetings of communities, tribes and castes. They may not lie outside the orbit of the state, and can be the locus for the mobilisation of tax revenue, the compulsory purchase of commodities, and the allocation of state welfare and information services (Ward *et al.*, 1978). The existence of a more or less regulated wholesale market for a given commodity indicates little about other aspects of marketing in that settlement, such as the degree of development of wholesale markets for other commodities or of retail trade, the co-existence of daily and periodic marketing, or the geographical role of a given wholesale marketplace in intra- and inter-regional systems of commodity flows (Jones, 1972, 1974).

Since capitalism penetrates peasant society irregularly it is sometimes possible to identify, as Blaikie *et al.* (1980), have achieved for Nepal, a form of circulation complementing a dominant mode of production. It is sometimes possible to observe a range of exchange and trading relations within one commodity market, or between commodity markets within a town, or between towns (see Harriss, 1981, 1984; Washbrook, 1973; Baker, 1984). As Chaudhuri (1979) observes in relation to 17th and 18th century India, the structure and organistion of buying and selling markets have differed widely from one society to another without impeding the free flow of goods and services between them though it has demanded considerable flexibility in commercial techniques. Such flexibility may amount to the extraction of surplus through various modalities, merchants being simultaneously "businessmen of the epoch of primitive accumulation and modern entrepreneurs ... (with) fingers in many pies, investing in trade, industry, finance and land" (Levkovsky, 1966, p.234). When a dominant process can be identified at the level of the wholesale market town, a distinction can be made between 'generative' and 'parasitic' urbanism. "A generative city will allocate a considerable amount of the surplus value accumulated within it to forms of investment that enlarge production ... in the city or in the surrounding rural area" (Harvey, 1973, p.233). A parasitic city will be the reverse, a sink of unproductive capital, its population the "excreta of a consumption system" rather than a production system (Chattopadhyay, 1969, p.221-22).

These are vivid but crude binary oppositions with which to end a demonstration of the great variety of ways in which peasant society can be commercialised. "Infinitely diverse combinations or elements of this or that type of capitalist evolution are possible" (Lenin, 1966, p.178), and "there cannot be any abstract generalisations about these conditions" (Kurien, 1977, p.430). An analysis of two village cases in Hausaland in semi-arid west Africa and two regional cases in

213

Tamil Nadu in semi-arid south India amply illustrate these statements, and it is to these that attention now turns.

The commercialisation of grain in Hausaland

Hausaland, straddling Northern Nigeria and Southern Niger, is a region of Sahelian West Africa with "a high degree of cultural, linguistic and religious uniformity" (Hill, 1972, p.xiv). Nevertheless, it also has the unusual characteristic for West Africa that "the population is so dense that farmers are obliged to farm all the cultivable land every year" (Hill, 1972, p.21).

The research of Raynaut (1973, 1975, 1977) has revealed interactions between exchange and production in a then deficit millet producing, but surplus groundnut producing, village near Maradi. Here, "cereals commerce does not indicate the existence of a (physical) surplus, nor does it assure better distribution between members of the community. Everything points to its origin in the financial vulnerability of certain family heads and it operates to accentuate their weakness and dependence, fuelled by their intense need for money" (Raynaut, 1973, p.34-35; my translation). The money for which goods are sold is needed to pay taxes in cash and this requirement alone amounted *on average* to 65 per cent of the value of the groundnut crop. It also forced the poorest half of the farming population to sell 35 per cent of their millet, most of which was locally redistributed.

In Hausa Niger very few individuals do not derive some income from trade, but for most people the intense circulation of goods and money does not offer the prospect of accumulation, money being immediately restored to the coffers of the State in the form of taxes (Mainet and Nicholas, 1964; Raynaut, 1975, vol.2, p.31; Raynaut, 1977, p.171). Because food has to be repurchased later, the terms of trade for farmers selling distress surpluses are automatically worse than for those surpluses which are not used for such purposes. For those who participate in the market economy on terms of distress, these mechanisms leave no surplus for money for investment in the limited new technology. Further mechanisms involving debt and an expansion of wage labour, though not here of total landlessness as yet, may be set in motion. The expansion of wage labour reduces time spent on the labourers' own lands, reduces their own production, increases distress sales and purchases, and the necessity to labour and to borrow money. Loans with interest (often 100 per cent and taken from seed) are commoner than loans without interest (Mainet and Nicholas, 1964, p.113; B.C.E.O.M., 1978, p.57; Raynaut, 1975, vol.2, p.37).

For the accumulators of grain and money "the benefits occurring from this intense mercantile activity fuel further commercial speculation, be it through purchasing vehicles for transport or be it by constructing houses for renting out. It is only rarely that such profits are directly invested in agriculture" (Raynaut, 1975, vol.2, p.33, my translation).

Linkages between marketing, finance and production have been described for Nigerian Hausaland in a case study of a millet and cotton growing village near Kaduna by Clough (1981, 1985). Here 30 per cent of farmers whose gross production does not meet their domestic food needs have to sell grain required for subsistence directly after harvest. They sell at low prices in order to meet the combined costs of taxes, debt repayment, house repair and ceremonies involving the purchase of manufactured goods against which the agricultural terms of trade are secularly deteriorating. In the pre-harvest rainy season these farmers work as seasonal paid wage labour on the farms of others in order to obtain food at double the post-harvest price, or they are forced to borrow money at seasonal interest rates of between 50 and 140 per cent often to be repaid in kind valued at post-harvest prices.

The successful middle farmers of Clough's studies have bimodal selling patterns, contributing to the post-harvest glut because of heavy social pressures for ceremonial or marriage expenditure. However, they tend to sell relatively less and (because of being better risks) borrow more than poor farmers. During the cultivation season grain will again be sold to raise money and to pay wage labour.

Large farmers are able to withold grain against mid-season price rises when they sell in order to pay wage labour and to invest in cattle, trade in which yields large speculative profits. The same large farmers link trade in cattle and trade in grain. The market is highly volatile at the micro-level, with prices offered to producers varying with the quantity of the individual lot sold, and therefore with the wealth of the seller.

Farmer-traders may also store grain in villages against off season price rises and lend out money borrowed in turn from inter-village specialised wholesalers. The latter carry out diversified longer distance, and frequently urban-based trade, and have higher average rates of return from the farmer-traders to whom they lend money. They operate in turn on money borrowed from a few urban wholesalers with whom they 'share profits' and whom they also simultaneously defraud. Here it is therefore possible to see a very clear pattern of money flows between the rural and urban areas.

Clough hypothesises that the social relations of exchange enable a small number of farmers to accumulate grain and money. The market structure is an hierarchical trading system financed by urban merchants and hereditary notables and tied by patron-client dependency relations. The system of linked money and grain markets leads to the accumulation of financial and physical surpluses through its indirect control of production. The linking of money and labour markets reinforces indebtedness and enables the extraction of a distress surplus. Distress sellers are forced to work later on as wage labour in order to obtain food whose earlier sale was equally forced on them. Hence, the system reduces their labour input on their own farms. Thus the market mechanism which commercialises the rural economy, in reducing returns to some farmers, constrains its own progress through society. Since production is constrained, profits from grain marketing come from exploiting supply/demand discrepancies through hoarding, evidenced by very large seasonal price swings (Hill, 1971, p.137; Watts and Shenton, 1978; Gore, 1978). Hoarding exacerbates seasonal supply shortages and reinforces the constraining mechanism on production.

These village studies in Hausaland clearly demonstrate four aspects of the commercialisation of grain:
(i) The existence of distress sale and pre-harvest buy-back (attested in Nigerian and Nigerien cases such that marketable surplus is less than that actually marketed) is a manifestation of superficial commercialisation which does not help develop

purchasing power on the home market (see Nadkarni, 1979, p.40).

(ii) This superficial commercialisation may be set in motion either by the social relations of exchange (as in Nigeria), in particular the need of some farmers to borrow money from other farmer-traders, or by the fiscal demands of the State (as in Niger).

(iii) Markets for both money and commodities are structured hierarchically in a way which permits the co-existence of large numbers of intermediaries divided between petty and monopoly scales of activity. The former may be highly financially dependent on the latter, rather than in the process of evolving into the latter.

(iv) The control by a small section of rural society over food and money markets also brings about the expansion of the market for labour which is at first disguised as kind repayments for loans.

It is now possible to turn to South India where all four types of market (land, labour, money and commodities) have existed for a long while and where the State has also intervened more profoundly in exchange than it has in Hausaland.

The Commercialisation of Grain in Tamil Nadu

Two contrasting regions will be considered. The first, 3,300 square kilometres comprising seven taluks of the eastern plain of North Arcot District, is a region of medium rainfall whose agricultural economy, dominated by high yielding rice and groundnuts, is based on water from tanks and energised open wells. The region is surplus in these grains. Here a class of capitalist farmers is emerging (Harriss, J., 1982a).

These farmers accumulate, are commercially oriented, use new technology, employ wage labour, and their productivity is greater than those of small peasants. However, the expansion of this capitalist class is constrained by the fact that further increases in production require investment in land. The land market has historically been very sticky because, on the one hand marginal petty producers "are being reproduced, albeit on unequal terms by merchants' and usurer's capital, and on the other hand their kinship systems maintain their land within a tight circle of kin" (Harriss, J., 1982a, p.436-7). So it is the finance and grain marketing system which here constrains the development of the land market apparently necessary to increase production, and which, in contrast to Hausaland, constrains the rate of expansion of the already very large landless labour market. A decade ago money was lent to small peasants by merchants at relatively low interest rates and these peasants were not discriminated against in terms of product price whatever their reasons for participation in market transactions. Despite or because of the fact that the State has begun to encroach on the financial role of the private merchant, land sales remain sticky, while the grain market continues to be crowded.

The mercantile sub-sector trading in grain comprises over 2,000 firms polarised between an oligopolistic elite and a large number of smaller merchants, the latter nevertheless mostly more wealthy and more landed than the average peasant producer. Insignificant quantities of grain are bulked from the network of periodic markets, which are important for petty general retailing and cattle wholesaling. Yet it is evident that the exchange of grain at periodic markets, is carried out by poorer, lower caste, often female traders, sellers and purchasers. It has the hallmarks of a socially and spatially separate commercial system (Harriss, 1976; Aiyasami and Bohle, 1981).

In the main, mercantile capital increasingly is generated not from agriculture but from private trading. Trading profits tend to be reinvested in trade, where the average annual rates of return exceed respectively those of private moneylending (the other activity invested in massively by traders), agro-processing industry, State moneylending and agricultural production (Harriss, 1981; Harriss and Harriss, 1984). Specific structural and behavioural conditions in the interlocked markets for money, inputs and products create imperfections where high average profits are possible from mercantile activity. A decade ago competition in money-lending at low interest to peasants by merchants was necessary to cement the particularistic social relations guaranteeing the supplies for speculation in the grain market. Today it is more complicated. Subsidised State credit is replacing that of the private trader. The merchant scours further afield, specialises, and lends to far flung agents who guarantee supplies. Low interest rates in the money market complement high returns from merchanting and cannot be dissociated from it.

The role of the State, through partial State trading by compulsory levies on rice merchants, results in a worsening of market imperfections which raise the average rates of return to private trade and attract entrants. Income tax laws encourage the splitting of joint firms and families into smaller nuclear units and thus enlarge the number of firms. The liberal supply of Indian government finance to inputs traders, who are frequently also products traders, in the form of loans on inventory and overdraft facilities swells the supply of private production credit. Subsidised production credit in the co-operative sector clearly has a restraining effect on private interest rates. Co-operative credit also compensates for private sector finance in localities where the latter is weakly developed. State financial institutions, in the forms of nationalised banks, co-operative banks and the financing done by the Department of Agriculture itself, also compete with each other and with private trade in the disbursement of production loans. Additionally, the early and mid-1970s saw the development of competing supplies of private production credit from pawnbrokers, 'finance companies' and even minor Government officials.

This system of finance and marketing feeds on commodity production and involves small producers compulsively in the market economy, but insofar as it ensures the survival of these small producers it simultaneously constrains the transformation of production relations. Thus in this case it also seems to constrain the pace of expansion of rice production. If production is being constrained relative to demand, the relatively large distributive margin fuels the system and attracts entrants (Harriss, 1981). This sector keeps up its high rates of profit relative to other sectors despite the proliferation of intermediaries. Average rates of profit conceal a combination of petty scale 'subsistence' trading and oligopolistic price manipulation helped by particular forms of State intervention.

Further research has emphasised the regional specificity of these complex relationships (Harriss, B., 1982). The last case study concerns 2,500 square kilometres of Avanashi and

Palladam Taluks, the driest part of Coimbatore District, one of the driest districts in Tamil Nadu, with a highly commercialised and diversified agriculture. This is based firstly on longstanding exploitation of a deepening water table with energised open wells and secondly on an unpredictable supply of rainfall, which makes rainfed agriculture very risky. Here there is a permanent deficit in grain production. The absence of technological innovations in millet and sorghum production makes for a slow growth rate in grain production in contrast to that of rice in North Arcot.

In this region and in striking contrast to North Arcot District, there are only about 150 grain merchants. Rice is mainly imported from the grain bowls of the State and the rice trade is massively polarised. On the one hand there are large scale long distance traders-cum-millers whose location tends to be urban. On the other hand there are rural petty traders in periodic markets often operating as disguised wage labour purchasing on credit to a clientele among the rural poor. Insofar as this 'consumption credit' is used to reproduce the recipient and his or her family, then it ought to be conceptualised as a special type of production credit for labour power. The millet trade is also subject to structural polarisation. The few large urban traders operate simultaneously as monopolistic controllers of local supplies for redistribution over time via their domination of storage facilities, and as dependent suppliers of raw materials for large scale animal feedstuffs compounding industries.

Major long term crisis conditions for grain production in the region have reinforced the polarisation of control of grain, and a chain of dependency of the rural destitute on large urban merchants via the precarious credit relations of petty food distributors.

Unlike the case of North Arcot District there is very little moneylending by traders for grain production. Such lending as occurs is not only petty in scale, it is short term in duration: when the risk attached to the harvest outcome is minimised. A ramification of the tendency to delink marketing from production finance is *not* that financial intermediaries are most specialised but instead that they are more diversified than in North Arcot.

Some kind of resolution to the long term crisis of grain production may be being found in rural industrialisation.

This development is not widely replicable over dryland India, and it derives its impetus from the specific conditions of the growth of the textile and engineering industries of the city of Coimbatore (Baker, 1984; Harriss, J., 1982b).

These case studies show how the operations of the food marketing system and its associated system of finance may constrain, or may reinforce other mechanisms which constrain, the pace of expansion of food production in semiarid agrarian economies. Additionally they may increase the vulnerability of the rural producers to food shortages, especially for those with no land of their own.

Commercialisation and the Nutrition of Peasant Households

The nutritional status of the peasant household is both the end and the beginning of the system of production and commercialisation for it is partly upon consumption that the health and productive capacity of a person depends. This in turn affects land use and costs of production (Maxwell, 1984). It is sometimes concluded that the commercialisation of agriculture results in malnutrition for a massive proportion of the existing peasant population (see Oculi, 1977; Wisner, 1976 for Kenya; George, 1985 more generally). It should be clear from this chapter that commercialisation does not affect all regions and classes of peasant society equally in terms of the household's ability to nourish and reproduce itself. Nevertheless, certain aspects of nutrition do need further clarification.

Malnutrition is hard to define and identify. If we confine ourselves to protein energy malnutrition (PEM), the increasingly low, varied and specific recommendations for calorie and protein intake (F.A.O. and W.H.O., 1973; Payne, 1976; F.A.O., 1977) by which PEM has to be measured, are subject to criticisms on the count of inappropriateness. Large numbers of the mildly and moderately malnourished are found apparently 'small but healthy' (Seckler, 1979; Lipton, 1983; Sukhatme, 1978, 1982), small because of the effects upon growth of diseases in infancy but not functionally or cognitively impaired in adulthood. Furthermore the new idea which explains the widely observed phenomenon of large intra-individual variations in the relationship between energy intake and output as resulting from complex metabolic adjustments, scotches altogether the concept of 'requirement' as a standard against which deviations may be measured, especially, as is frequently done for peasant societies, using nationally aggregated food budgeting data (Pacey and Payne, 1985). This has radical implications. Even the common inferences about unequal allocation of food within peasant households (Kynch and Sen, 1985) need to be considered cautiously in view of suspicions that highly specified food intake standards, as of children under five, could be overestimates, and that existing field techniques to establish energy intake are vulnerable to large measurement errors (Wheeler, 1985). Notwithstanding, it has been possible to build up from a patchy literature scenarios for Subsaharan Africa and Southern Asian peasant societies which elaborate the relationship between commercialised food production and malnutrition and can be evaluated in the context of the case studies already summarised in this chapter. Given the paucity of relevant research these are best regarded at present as hypotheses for detailed field testing.

The Sub-Saharan African scenario

We have seen that in sub-Saharan Africa land, labour and grain are not as highly commercialised as in South Asia. However, there is evidence that the evolution of land tenure is reducing the security of women to usufructuary rights over cultivable land (Dey, 1982; Trenchard, 1987). Likewise, the physical degradation of land through soil erosion and desertification increases the time-burden of reproductive tasks such as the hunt for water and fuel, also thought to be the preserve of women. Although sorghum and millet tend to be poorly commercialised, rice and maize have both been subject to transformations in production technology and rapid commercialisation (Smith *et al.*, 1981; van der Laan *et al.*, 1985). In this process male farmers are thought to be increasing their control over commercialised production and the income accruing from it, while women are increasingly responsible for household provisioning (Whitehead, 1981; Carr, 1984). Where labour has been commercialised it has taken the form of wage work in plantations or mines necessitating the divisive migration of adult men. In the female headed households which remain, productive and reproductive work burdens are intensified. There may be less oportunity for women to trade, where income from commerce is closely and positively associated with the nutritional status of the children of the female trader (Trenchard,1987). Such complex forms of commercialisation of labour and commodities are thought to increase women's work burdens more than those of men, with malevolent implications for reduced birth weights and growth rates of children, and reduced lactational capacities for their mothers. There is evidence for seasonality in births, and for peasants to be lactating in precisely the preharvest rainy season when the energy requirements for agriculture and household maintenance exceed energy intakes from depleted grain stores and when debilitating infections abound (Chambers *et al.*, 1981; Bleiburg *et al.*, 1980; Brun *et al.*, 1981; Hill, 1985). The nutritional significance of such evidence is debated. Seasonal fluctuations in weight may be an efficient nutritional adaptation to the agrarian ecology: better to store energy in the body than rely on uncertain grain markets. It may be physiologically more efficient, seasonal energy surpluses and deficits enabling the peasant to be taller than s/he would be if s/he had eaten exactly according to need all year round, and better physiologically prepared for the sharp peaking of energy expenditure in agricultural production.

South Asian scenario

In South Asia massive inequalities in income and assets are reflected in per caput food intakes (Lipton, 1983). Two thirds of the entire world's malnutrition is here. Estimates of the proportion of the rural population suffering malnutrition range from 15 per cent (Sukhatme, 1981) to 50 per cent (Dandekar, 1981). Those households which are assetless or which, like weavers and fishers, have to sell what they produce in order to buy grain are particularly nutritionally vulnerable (Appadurai, 1984). Alongside a process where dependence upon labour or other commodity markets is associated with lack of access or entitlement to food, runs another process - that of disenfranchisement - wherein certain household members systematically exclude themselves or are excluded from equal access to food. If we leave aside the problematic food intake data, we are nonetheless left with medical and anthropometric evidence of nutritional discrimination against females and against children under five in the north of the sub-continent (Chen *et al.*, 1981; Sen, 1984). In addition the male may be favoured in breast-feeding (Levinson, 1972), in priority in eating (Appadurai, 1981), in consumption of prestige food (Sharma, 1983) and in control of the household food budget or basket (Harriss, 1986). The disenfranchisement of women affects them at times in their lives when they are most nutritionally vulnerable, during infancy and reproductive years, and when the household itself is most nutritionally vulnerable, at the time when the relationship between dependents and workers is maximised.

There is some north Indian evidence that in households with the most precarious entitlements to grain, the disenfranchisement of female children to grain, is most intense. A given level of discrimination against females may be most fatal among the poorest, landless and marginally landed peasants, and the lowest castes, who are likely to be undergoing distress commercialisation of labour and commodities (Levinson, 1972; Sen and Sengupta, 1983; Dasgupta, 1987; Wadley and Derr, 1987). However, even in the north new evidence conflicts. The Jefferys (1987) have found sex bias maximised both among wealthy landed Muslim households and among poor assetless harijans, in villages in Uttar Pradesh. Warrier (1987) records a complete absence of sex bias among assetless pauperised tribals in West Bengal. It is impossible to explain these manifestations simply with reference to the commercialisation of female labour and female economic status. Nor is it a matter of cultural versus material determination. Rather, the social geography of sex bias depends on the interaction between specific features of kinship, culture, the position of the household in the labour process, and the form of commercialisation of male and female labour and of products (Harriss, 1987; Harriss and Watson, 1987).

Explanations involving commercialisation in this long commercialised economy tend to be banal or arcane and are hampered by lack of empirical evidence. Discrimination could result from low female status derived from low participation rates in wage work (Miller, 1981). Certainly women earn lower wages and the nutritional status of male peasants has been found to relate closely to earnings from wage work (Ryan, 1982). Generally the proportion of women in the commercialised labour force has been declining since 1901 (Omvedt, 1978; Mencher, 1980). Instances where the labour force is being feminised in response to the impact of new technology upon the local gender division of tasks are swamped by the massive displacement of women by mechanisation in production and in post-harvest processing (Mukherjee, 1983; Harriss and Kelly, 1982), and by the voluntary withdrawal of female labour in richer households engaged in normal commercialisation (Agarwal, 1984). This emphasis on participation has de-emphasised the uncommercialised work of women both in household production and in reproduction, the significance of which for their enfranchisement to food is still debated. North-south regional differences in enfranchisement to food have also been related to variations in female autonomy expressed in property rights and marriage transactions (Dyson and Moore, 1983), but the ethnographic evidence for less autonomy in the north than in the south is still not convincing.

Commercialisation is bringing about alterations in the gender division of labour in these peasant societies which tend to reduce the enfranchisement to food of precisely those who are most nutritionally vulnerable in physiological and social terms. The modalities by which the poor peasantry gains access to grain may become more complex, depending on the relationships between the price of grain, that of their labour, and those of other commodities produced by them (Sen, 1981). Seasonal variations in grain entitlement may be sudden and large (Cutler, 1985 for Bangladesh), even under circumstances of adequate supply and competitive marketing. But the least-entitled may, at the best of times, face different exchange relations not reflecting competitive forces, and reasonably competitive markets may become distorted or oligopolistic at the worst of times (Seaman and Holt, 1980; Spitz, 1977, 1985), in scarcity, precisely when the dependence of peasant societies upon the market is at a maximum.

Conclusion

Having surveyed several of the theoretical arguments concerning commercialisation and market exchange, this chapter has provided a detailed assessment of the differential influences of grain commercialisation in two contrasting areas of Africa and Asia, and explored its implications for nutritional levels in peasant households. Three broad conclusions concerning the nature of rural-urban resource transfers emerge.

The first is that although many of the broad generalisations that can be made concerning urban-rural interaction do appear to be appropriate at the macro level, they can nevertheless conceal important local differences in the ways in which commercialisation has been associated with increased inequality between individuals and classes in society. Along with this, as the discussion of nutrition levels has indicated, there is a pressing need for more detailed local empirical research on, for example, the reasons for the differential levels of nutrition between men and women, and between different social classes.

That having been said, this chapter has indicated that both in Hausaland and in Tamil Nadu the commercialisation process has given rise to significant transfers of grain, money and power from rural to urban areas. The precise mechanisms by which this has taken place, however, are highly complex, and simplistic explanations which only see urban-based merchants as exploiting a rural peasantry must be discounted. Both in Hausaland and in Tamil Nadu there are very few individuals who do not derive some income from trade. Landless labourers must thus resell part of the kind rewards of their own direct labour in order to pay Poll Taxes, repay debts and consume non-kind items. Periodic markets in both areas are likewise regularly full of poor peasant producers selling part of their produce in order to pay off debts, purchase inputs, and obtain other sources of food. It is nevertheless the rich farmers, who have access to the urban economic and political networks, and the urban-based merchants and wholesalers themselves, who through their control of the marketing system are able to extract and redirect any rural surpluses that are created. The periodic market systems are focused through commodity flows on larger, often daily, markets located in the main urban centres of population. The inability of most poor rural peasants to transport their produce to these towns, and to gain access to the marketing facilities and exchange relations therein, thus precludes them from the opportunity of accruing the profits available to the mercantile firms and smaller merchants.

Finally, as was emphasised particularly in the case study of Tamil Nadu, trading profits tend to be reinvested not in rural areas but in further commercial speculation. Financial intermediation comes to play an increasingly important role in the lives of the more affluent farmer-traders and merchants. While it is difficult to consider this financial activity as uniquely urban, the focus through which the grain and money flows is indeed urban. Much of the surplus extracted from rural areas is thus unavailable for reinvestment in agriculture, though this chapter has shown that it is not production *technology* but rather the way in which production and distribution are organised, both at the level of the market and at the micro level within the individual household that conditions well being.

Rural-Urban Grain and Resource Transfers

REFERENCES

Adnam, S. (1985) 'Classical and contemporary approaches to agrarian capitalism', *Economic and Political Weekly* Review of Political Economy XX(30), PE53-64

Agarwal, B. (1984) 'Rural women and High Yielding Variety rice technology', *Economic and Political Weekly* Review of Agriculture, March, 19, 13

Aiyasami, U. and Bohle, H.G. (1981) 'Market access as constraint on marginal and small farmers', *Economic and Political Weekly* Review of Agriculture XVI(13), A29-36

Appadurai, A. (1981) 'Gastropolitics in South Asia', *American Ethnologist* 8, 494-511

Appadurai, A. (1984) 'How moral is South Asia's economy? - a review article', *Journal of Asian Studies* 43, 481-97

Baker, C.J. (1984) *An Indian Rural Economy 1880-1955: The Tamiland Countryside*, Oxford University Press, Delhi

Bardhan, P. and Rudra, A. (1978) 'Interlinkage of land, labour and credit relations', *Economic and Political Weekly* XIII(8), 367-84

B.C.E.O.M. (1978) *Project de Routes Rurales en Republique du Niger: Etude de la Commercialisation et du Stockage des Produits Argicoles.* Ministère des Travaux Publics, des Transports et de l'Urbanisme, Niamey, Niger

Bharadwaj, K. (1974) *Production conditions in Indian Agriculture*, Cambridge University Press, London

Bharadwaj, K. (1985) 'A view on commercialisation in Indian agriculture and the development of capitalism', *Journal of Peasant Studies* 12, 7-25

Bhaduri, A. (1977) 'On the formation of usurious interest rates in backward agriculture', *Cambridge Journal of Economics* 1, 341-52

Bhaduri, A. and Sud, B. (1981) 'Class relations and commercialisation in Indian agriculture', paper for seminar on Commercialisation in Indian Agriculture, Centre for Development Studies, Trivandrum, mimeo.

Blaikie, P.M., Cameron, J. and Seddon, J.D. (1980) *Nepal in Crisis: Growth and Stagnation at the Periphery*, Clarendon, Oxford

Bleiburg, F.M., Brun, T.A. and Goiham, S. (1980) 'Duration of activities and energy expenditure of female farmers in dry and rainy seasons in Upper Volta', *British Journal of Nutrition* 43, 71

Bottomley, A. (1964) 'Monopoly profit as a determinant of interest rates in underdeveloped rural areas', *Oxford Economic Papers* 16, 431-37

Bressler, R.G. Jr and King, R.A. (1970) *Markets, Prices and Interregional Trade*, John Wiley, New York

Bromley, R.J. (1974) *Periodic Markets, Daily Markets and Fairs: A Bibliography*, Centre for Development Studies, University College of Swansea, U.K.

Bromley, R.J. (1979) *Periodic Markets, Daily Markets and Fairs: A Bibliography Supplement to 1974*, Centre for Development Studies, University College of Swansea, U.K.

Brun, T., F. Bleiburg and Samuel, G. (1981) 'Energy expenditure of male farmers in dry and rainy seasons in Upper Volta', *British Journal of Nutrition* 45, 67

Byres, T.J. (1972) 'Industrialisation, the peasantry and the economic debate in post Independence India, in: Bhuleshkar, A.V. (ed.) *Towards the Socialist Transformation in the Indian Economy*, Progress Bombay, 223-47

Byres, T.J. (1974) 'Land reform, industrialisation and the marketed surplus in India: an essay in the power of rural bias, in: Lehmann, D. (ed.) *Agrarian Reform and Agrarian Reformism*, Faber, London, 221-61

Carr, M. (1984) *Blacksmith, Baker, Roofing Sheet Maker: Employment for Rural Women in Developing Countries*, Intermediate Technology Group Publications, London.

Chambers, R., Longhurst, R. and Pacey, A. (eds.) (1981) *Seasonal dimensions to Rural Poverty*, Frances Pinter, London

Chattopadhyay, B. (1969) 'Marx and India's Crisis', in: Joshi, P.C. (ed.) *Homage to Karl Marx*, Peoples' Publishing House, Delhi, 205-60

Chaudhuri, K.N. (1979) 'Markets and traders in India during the 17th and 18th centuries', in: Chaudhuri, K.N. and Dewey, C.J. (eds.) *Economy and Society:*

Essays in Indian Economic and Social History, Oxford University Press, Delhi, 143-62
Chen, L.C., Huq, E. and D'Souza, S. (1981) 'Sex bias in the family allocation of food and health care in rural Bangladesh', *Population and Development Review* 1, 55-70
Clough, P. (1981) 'Farmers and traders in Hausaland', *Development and Change* 12, 273-92
Clough, P. (1985) 'The social relations of grain marketing in Northern Nigeria', *Review of African Political Economy* 34, 16-34
Cutler, P. (1985) 'Detecting food emergencies: lessons from the 1979 Bangladesh crisis', *Food Policy*, August, 207-24
Dasgupta, M. (1987) 'The second daughter: neglect of female children in rural Punjab, India', paper for workshop on Differential Female Mortality and Health Care in South Asia, American S.S.R.C., Dhaka
Dandekar, V.M. (1981) 'On measurement of poverty', *Economic and Political Weekly* XVI(30), 1241-50
Deere, C.D. (1979) 'Rural women's subsistence production in the capitalist periphery', in: Cohen, R., Gutkind, P.C.W. and Brazier, P. (eds.) *Peasants and Proletarians*, Hutchinson, London
Deere, C.D. and de Janvry, A. (1979) 'A conceptual framework for the empirical analysis of peasants', *American Journal of Agricultural Economics*, November, 601-11
Dey, J. (1982) 'Development planning in the Gambia: the gap between planners' and farmers' perceptions, expectations and objectives', *World Development* 10, 377-96
Dyson, T. and Moore, M.P. (1983) 'Kinship structure, female autonomy and demographic behaviour: regional contrasts within India', *Population and Development Review* 9, 35-60
F.A.O. (1977) *Fourth World Food Survey*, Food and Agriculture Organization, Rome
F.A.O. and W.H.O. (1973) 'Energy and protein requirements: report of a joint FAO/WHO ad hoc expert committee', *World Health Organisation Technical Report* No.522, World Health Organisation, Geneva
Friedmann, H. (1983) 'The political economy of food: the rise and fall of the post war international food order', *American Journal of Sociology* 88, 248-86
George, S. (1985) *Ill Fares the Land: Essays on Food and Power*, Readers and Writers Publishing Co-operative, London
Ghatak, S. (1975) 'Rural interest rates in the Indian economy', *Journal of Development Studies* 11, 190-201
Gore, C. (1978) 'The terms of trade of food producers as a mechanism of rural differentiation', *Bulletin of the Institute of Development Studies*, Brighton, 20-24
Gormsen, E. (ed.) (1976) *Market Distribution Systems*, Mainzer Geographische Studien Heft 10, Mainz, Germany
Guillet, D. (1981) 'Surplus extraction, risk management and economic change among Peruvian peasants', *Journal of Development Studies* 18, 1-24
Harriss, B. (1976) 'The Indian ideology of growth centres', *Area* 8, 263-69
Harriss, B. (1980) 'There's a method in my madness - or is it vice-versa? Measuring agricultural market performance', *Food Research Institute, Stanford Studies* XVI, 40-56
Harriss, B. (1981) *Transitional Trade and Rural Development*, Vikas, New Delhi
Harriss, B. (1982) 'Production and circulation of cereals and money in a dryland tract of Coimbatore District', in Pradham, M. (ed.) *Small Farmer Development and Credit Policy*, Agricultural Credit Training Institute, Kathmandu, Nepal, 265-81
Harriss, B. (1984) *State and Market*, Concept, Delhi
Harriss, B. et al. (1984) *Exchange Relations and Poverty in Dryland Agriculture*, Concept, Delhi
Harriss, B. (1986) *Meals and Noon Meals in South India*, Occasional Papers 31, School of Development Studies, University of East Anglia, Norwich

Harriss, B. (1987) 'Differential female mortality and health care in South Asia', report to the Ford Foundation, New York

Harriss, B. and Harriss, J. (1984) "Generative' or 'Parasitic' urbanism? Some observations from the recent history of a South Indian market town', in: Harriss, J. and Moore, M. (eds.) *Development and the Rural Urban Divide*, Frank Cass, London, 82-101

Harriss, B. and Kelly, C. (1982) 'Food processing: policy for rice and oil technology in South Asia', *Bulletin Institute of Development Studies* 13, 32-44

Harriss, B. and Watson, E. (1987) 'The sex ratio in South Asia', in: Momsen, J. and Townsend, J. (eds.) *Gender and Geography in the Third World*, Hutchinson, London

Harriss, J. (1982a) *Capitalism and peasant farming*, Oxford University Press, Bombay

Harriss, J. (1982b) 'Character of an urban economy: small scale production and labour markets in Coimbatore', *Economic and Political Weekly* XVII(23), 945-54

Harriss, J. and M. Moore (eds.) (1984) *Development and the Rural Urban Divide*, Frank Cass, London

Harvey, D. (1973) *Social Justice and the City*, Edward Arnold, London

Hill, A. (ed.) (1985) *Population, Health and Nutrition in the Sahel*, Kegan Paul International, London

Hill, P. (1969) 'Hidden trade in Hausaland', *Man* 4, 392-409

Hill, P. (1971) 'Two types of West African house trade', in: Meillassoux, C. (ed.), *The Development of Indigenous Trade and Markets in West Africa*, International African Institute, Oxford University Press, London, 303-10

Hill, P. (1972) *Rural Hausa: A Village and a Setting*, Cambridge University Press, London

Jeffery, P., Jeffery, B. and Lyon, A. (1987) 'Domestic politics and sex differences in mortality: a view from rural Bijnor District, U.P.', paper for workshop on Differential Female Mortality and Health Care in South Asia, American S.S.R.C., Dhaka

Jones, W.O. (1972) *Marketing Staple Food Crops in Tropical Africa*, Cornell, University Press Ithaca, New York

Jones, W.O. (1974) 'Regional analysis and agricultural marketing research in tropical Africa: concepts and experience', *Food Research Institute Studies* 13, 3-28

Kurien, C.T. (1977) 'Abstract generalisations', *Economic and Political Weekly* 10, 428-30

Kynch, J. and Sen, A.K. (1983) 'Indian women: well being and survival', *Cambridge Journal of Economics* 7

van der Laan, L., Arhin, K. and Hesp, P. (1985) *Marketing Boards in Tropical Africa*, African Studies Centre, Leiden, Netherlands

Lenin, V.I. (1960) *The Development of Capitalism in Russia*, Foreign Language Publishing House, Moscow

Levinson, F.J. (1972) *An Economic Analysis of Malnutrition Among Young Children in Rural India*, MIT/Cornell University Press, Ithaca

Levkovsky, I. (1966) *Capitalism in India - Basic Trends in its Development*, People's Publishing House, Bombay

Lipton, M. (1977) *Why Poor People Stay Poor: Urban Bias in World Development*, Temple Smith, London

Lipton, M. (1983) *Poverty Hunger and Undernutrition*, World Bank Staff Working Paper, Washington

Mainet, G. and Nicholas, G. (1964) *La vallée du Gulbi de Maradi: enquête socio-économique*, Documents des Etudes Nigériennes No.16, IFAN/CMRS Niamey, Niger

Marx, K. (1974) *Capital: A Critical Analysis of Capitalist Production*, Lawrence and Wishart, London

Maxwell, S. (1984) 'Health, nutrition and agriculture: linkages in farming systems research', Discussion Paper No.195, Institute of Development Studies, Sussex

McNeill, G. (1986) Energy undernutrition in adults in rural South India, unpublished Ph.D thesis, London School of Hygiene and Tropical Medicine

Mencher, J. (1980) 'The lessons and non lessons of Kerala: agricultural labour and poverty', *Economic and Political Weekly* XV, 1781-1802

Michie, B.H. (1978) 'Baniyas in the Indian agrarian economy: a case of stagnant entrepreneurship', *Journal of Asian Studies* 37, 637-52

Miller, B. (1981) *The Endangered Sex: Neglect of Female Children in Rural North India*, Cornell University Press, Ithaca, New York

Minvieille, J.P. (1976) *Migrations et Economies Villageoises dans la Vallée du Senegal*, O.R.S.T.O.M., Dakar, Senegal

Mishra, S.C. (1981) Patterns of long run agrarian change in Bombay and Punjab, unpublished Ph.D. thesis, Cambridge University

Mukherjee, M. (1983) 'The impact of modernisation on women's occupations: a case study of the rice husking industry in Bengal', *Indian Economic and Social History Review* 20, 1

Nadkarni, M.V. (1979) 'Marketable surplus, market dependence and economic development', *Social Scientist* 83, 35-50

Nash, M. (1967) 'Indian economies', in: *Handbook of Middle Indians Series vol.6, Social Anthropology*, University of Texas Press, Austin

Nisbet, C. (1967) 'Interest rates and imperfect competition in the informal credit market of rural Chile', *Economic Development and Cultural Change* 16, 73-90

Oculi, O. (1977) 'Colonial capitalism and malnutrition: Nigeria, Kenya and Jamaica', unpublished Ph.D. thesis, Wisconsin

Omvedt, G. (1978) 'Women and rural revolt in India', *Journal of Peasant Studies* 7, 2

von Oppen, M. (1985) 'Agricultural marketing and aggregate productivity', in: I.C.R.I.S.A.T. *Agricultural Marketing in the Semi Arid Tropics*, I.C.R.I.S.A.T., Hyderabad

Pacey, A. and Payne, P.R. (1985) *Agricultural Development and Nutrition*, Hutchinson, London

Pouchepadass, J. (1981) 'Peasant economy and the market system in early 20th century Bihar', School of Oriental and African Studies, Conference on modern South Asian Studies, London, mimeo

Prasad, P.H., Rodgers, G.B., Gupta, S., Sharma, A.N. and Sharma, B. (1986) 'The pattern of poverty in Bihar', Population and Labour Policies Working Paper 152, World Employment Programme, International Labour Office, Geneva

Rao, C.H.H. and Subbarao, K. (1976) 'Marketing of rice in India: an analysis of the impact of producers' prices on small farmers', *Indian Journal of Agricultural Economics* 31, 1-15

Rao, G.N. (1985) 'Transition from subsistence to commercialised agriculture: Krishna District of Andhra, 1850-1900', *Economic and Political Weekly* XX(25-6), 60-9

Raynaut, C. (1973) *La Circulation Marchande des Céréales et les Mécanismes d'Inegalité Economique*, Université de Bordeaux, Centre D'Etudes et de Recherches Ethnologiques, Cahier no.2

Raynaut, C. (1975) 'Le cas de la région de Maradi (Niger)', in: Copans, J. (ed.), *Famines et Secheresses en Afrique*, Maspero, Paris, vol.2, 5-43

Raynaut, C. (1977) 'Circulation monétaire et évolution des structures socio-économiques chez les Haoussas du Niger', *Africa* 47, 160-71

Ryan, J. (1982) *Progress Report No.38*, I.C.R.I.S.A.T., Hyderabad, India

Sarkar, S. (1979) 'The marketing of foodgrains and patterns of exploitation', Department of Economics Occasional paper No.1, Visva Bharathi, Santiniketan, India

Sarkar, S. (1981) 'Marketing of foodgrains: an analysis of village survey data from West Bengal and Bihar', *Economic and Political Weekly*, September, Review of Agriculture XVI(39), A103-8

Seaman, J. and Holt, J. (1980) 'Markets and famines in the Third World', *Disasters* 4, 283-97

Seckler, D. (1979) 'Small but healthy: a crucial hypothesis in the theory, measurement and policy of malnutrition', Ford Foundation, New Delhi, mimeo

Sen, A. (1981) *Poverty and Famines: An Essay in Entitlement and Deprivation*, Clarendon, Oxford

Sen, A.K. (1984) *Resources, Values and Development*, Blackwell, London

Sen, A.K. and Sengupta, S. (1983) 'Malnutrition and the sex bias', *Economic and Political Weekly* 18, 855-64
Sharma, S. (1983) Food distribution patterns in farm families, unpublished M.Sc. thesis, Sukhadia University, Udaipur, India
Smith, C.A. (ed.) (1976) *Regional analysis: Volume I Economic Systems, Volume II Social Systems*, Academic Press, New York
Smith, R.H.T. (ed.) (1978) *Market Place Trade: Periodic Markets, Hawkers and Traders in Africa, Asia and Latin America*, University of British Columbia, Vancouver, Canada.
Smith, V.E. *et al.* (1981) 'Development and Food Consumption Patterns in Rural Sierra Leone', *Food and Nutrition* 7, 24-32
Spitz, P. (1977) 'Silent violence: famine and inequality', U.N.R.I.S.D., Geneva, mimeo.
Spitz, P. (1985) 'Drought and self provisioning', U.N.R.I.S.D., Geneva, mimeo.
Sukhatme, P.V. (1978) 'Assessment of adequacy of diets at different income levels', *Economic and Political Weekly* XIII, Special Number, August, 1373-84
Sukhatme, P.V. (1981) 'Measuring the incidence of malnutrition: a comment', *Economic and Political Weekly* XVI, 1034-36
Sukhatme, P.V. (1982) 'Relationship between malnutrition and poverty', in *Proceedings of the First National Conference on Social Sciences*, 1981, ICSSR, New Delhi
Tax, S. (1953) *Penny Capitalism: a Guatemalan Indian Economy*, Publication 16, Smithsonian Institution, Washington D.C.
Trenchard, E. (1987) 'Women's work, health and nutrition in subsaharan Africa', in: Momsen, J. and Townsend, J. (eds.) *Gender and Geography in the Third World*, Hutchinson, London
Wanmali, S. (1981) *Periodic Markets and Rural Development in India*, B.R. Publishing Corporation, Delhi
Ward, R.G. *et al.* (1978) 'Market Raun: the introduction of periodic markets to Papua New Guinea', in: Smith, R.H.T. (ed.) *Regional Analysis: Volume I Economic Systems; Volume II Social Systems*, Academic Press, New York
Warrier, S. (1987) 'Daughter disfavour, women's work and autonomy in rural West Bengal', paper for the Workshop on Differential Female Mortality and Health Care in South Asia, American S.S.R.C., Dhaka
Washbrook, D. (1973) 'Country politics: Madras 1880 to 1930', *Modern Asian Studies* 7, 475-531
Watson, E. and Harriss, B. (1984) 'Discrimination on money and commodity markets', in: Harriss, B. *et al.*, *Exchange Relations and Poverty in Dryland Agriculture*, Concept, New Delhi
Watts, M. and Shenton (1978) 'Capitalism and hunger in northern Nigeria', *Review of African Political Economy* 13
Wheeler, E. (1985) 'Intra household food allocation: a review of evidence', London School of Hygiene and Tropical Medicine, London, mimeo.
Whitehead, A. (1981) 'I'm hungry mum: the politics of domestic budgeting', in: Young, K., Wolkowitz, C. and McCullagh, R. (eds.) *Of Marriage and the Market*, CSE Books, London
Wisner, B. (1976) 'Man-made famine in Eastern Kenya: the interrelationship of environment and development', Discussion Paper 96, Institute of Development Studies, Brighton
Young, K., Wolkowitz, C. and McCullagh, R. (eds.) (1981) *Of Marriage and the Market*, CSE Books, London

8

URBAN-RURAL RELATIONS AND THE COUNTER-REVOLUTION IN DEVELOPMENT THEORY AND POLICY

STUART CORBRIDGE

Despite the continuing ascendency of Marxian and post-Marxian theories in 'development geography' (Smith, 1984; Forbes, 1984; Corbridge, 1986; Peet, 1987), it is apparent that a counter-revolution is afoot in development economics. Slowly through the 1970s, and with vigour and confidence in the 1980s, a group of scholars has emerged to challenge "Keynes and neo-Keynesianism, 'structural' theories of development and the use of economic planning for development purposes" (Toye, 1987, p.vii). In their stead, the counter-revolutionaries advocate a world of rolled-back governments and public sectors, with competitive free markets left to sponsor effective human capital formation, resource allocation and growth. Central to this project is a reassessment of the structure of urban-rural relations in the developing world. Counter-revolutionaries identify a Third World beset by urban bias, by excessive rural-to-urban migration and by inefficient import-substituting industries. All of these deficiencies they blame upon a legacy of Keynesian 'duoeconomics'. For their part, the counter-revolutionaries urge Third World countries to embrace the economics of comparative advantage and subjective preference. It is in the prescriptions of neo-classical economics that the Third World will find its route to 'accelerated development' (World Bank, 1981).

This essay takes seriously the political appeal of the counter-revolution. It also considers the logic of neo-classical economics as it informs a study of urban-rural relations. The paper comprises three main sections. The first section examines the alleged connections between Keynesian economics and the promotion of urban-industrial growth in the developing world. The second section outlines the critique of this 'orthodox' economics as it is developed by leading counter-revolutionaries. Attention is focused on the claims of monoeconomics versus duoeconomics, on the importance of market price signals and on the cumulative impact of 'political distortions' in the constitution of urban-rural relations. A review of the neo-classical resurgence in development studies is presented in the third section. The review is neither dismissive nor laudatory. Certain strengths of the counter-revolutionary discourse are noted and are defended against the simpler critiques of neo-classical economics as static equilibrium analysis. Nevertheless, the counter-revolution is found wanting in its black-box approach to the political conditions of existence of economic structures and agency. The paper also rejects the privileged position assigned to *the market* in the construction of the space-economies of the developing world. Real markets exist in space and in time, and are constituted according to varying principles and power relations. To promote as panacea an abstract 'market' is to conceal the necessary imperfections and inequalities of particular economic systems.

Dirigisme, statism and "orthodox" development economics

It is a mistake to suppose that the leading counter-revolutionaries in development studies (Balassa, 1985; Bauer, 1981, 1984; Beenstock, 1983; Clayton, 1983; Haberler, 1987; Johnson, 1971; Krueger, 1978; Lal, 1983; Little, 1982; Schultz, 1987) are united in their diagnoses of the 'development malaise'. John Toye (1987, p.72) is right to distinguish between the continental liberalism of a Peter Bauer and the Benthamite utilitarianism of Balassa, Lal and Little. An antipathy to Keynes and to Keynesianism is also more evident in the work of Harry Johnson than it is in the work of Schultz and Haberler, each of whom reclaims the dying Keynes from his erstwhile followers. Nevertheless, there is a common thread to the counter-revolution's portrayal of the genesis, and subsequent malformation, of development studies. Three themes are brought repeatedly to the reader's attention.

A first theme concerns the birth of development economics in the wake of the Great Depression of the 1930s. According to Gottfried Haberler (1987, p.53), "When the problems of development were thrust upon economists by the break-up of colonial empires in Asia and Africa during the Second World War and shortly thereafter, faith in liberalism, in free markets, and in free enterprise was probably at its lowest point since the early nineteenth century. No wonder that the stance of much of development economics, too, was far from liberal". Haberler continues, "The story of the decline of liberalism begins with World War One, 1914-18. This war marked the end, or the beginning of the end, of an epoch - the epoch of liberalism, of relatively free trade, of the gold standard, of free migration, free travel without a passport among most countries (excluding Russia but including the United States)" (Haberler, 1987, p.54). Although the world economy recovered in the 1920s, Haberler contends that the tariff walls put in place during the Great War allowed for further and more sweeping protectionism in the early 1930s. This, in turn, set the stage for the critique of liberal economics associated with Keynes and the Cambridge School.

We will return later to Haberler's revisionist interpretation of the Great Depression. For the moment it is enough to note that the Depression offered a lesson to development studies. The lesson concerns the relative values of planning and the market mechanism. The Depression is said to have encouraged the followers of Keynes - and Prebisch (1961) especially - to see in mature capitalism a necessary tendency to secular stagnation. Says Haberler (1987, p.56), "The general picture underlying the Keynesian policy prescriptions was that of a 'mature' economy that is subject to more or less continual deflationary pressure, chronic overspending and a scarcity of investment opportunities because of a slowdown of technological progress". By contrast, the economies of Germany and the USSR gained prestige in the 1930s. Deficit spending in Hitler's Germany, together with collectivisation and planned industrialisation in the USSR, seemingly had worked to shelter the 'totalitarian' economies from "the depression that engulfed the capitalist West" (Haberler, 1987, p.55).

The contrast between market failure and planned socialist growth stamped an emergent development studies with its second characteristic: an attachment to 'dueeconomics'. According to Albert Hirschman (1982) there are two types of development economist. On the one hand are the followers of 'monoeconomics'. These men and women believe that "standard economic theory is as applicable to the scarcity problems that confront low-income countries as to the corresponding problems of high-income countries" (Schultz, 1987, p.23). Markets function in the so-called Third World much as they do in the First World, albeit with higher levels of uncertainty. Economic agents in the Third World take decisions according to the logic of their First World counterparts. They seek to maximise their utilities, they seek an optimium allocation of resources and they exhibit the quality of entrepreneurship. Above all, farmers and industrialists in the Third World respond to the structures of incentives set either (and rightly) by the economic market, or (and wrongly) by the political market.

On the other hand is duoeconomics. According to the Keynesian revolution of the 1940s and 1950s the teachings of market economics are of little relevance in the developing world. A new ('duo') economics has to be fashioned which will recognise the particular and defining conditions of underdevelopment, namely: (i) the persistence of massive unemployment (open and/or disguised); (ii) the widespread existence of market failure, on account both of underdeveloped transport and communications systems and of poorly developed financial and economic networks; (iii) the constraints of culture and tradition which incline economic agents to behave other than as utility maximisers; and (iv) a tendency to a secular deterioration in the terms of trade. This last condition assumes a singular importance for the critics of duoeconomics. Just as development economics emerged in the post-Depression years, so too it took shape amidst a slump in the prices of internationally traded commodities. The combination had predictable results. An emerging duoeconomics concluded that industrialisation and the export of maufactured goods was a precondition of development. It also favoured state action both to direct resources to the urban-industrial sector and to protect local manufacturing enterprises from competing industrial imports.

The broad contours of a duoeconomics committed to import-substitution industrialisation are etched, thirdly, into a series of developmental models which internalise this basic logic. A first set of models concerns the 'big-push'. Leading counter-revolutionaries note the commitment of duoeconomics to a catastrophist interpretation of the development process. The Rostow model (Rostow, 1960) offers a widely known version of this thesis and it faithfully rehearses the theoretical and policy insights of its genre. The Rostow model is committed both to Westernisation and to industrialisation. For Rostow, "the essential condition or the central problem of economic development [is] to raise the saving and investment ratios of the underdeveloped countries above 10-12 per cent of their national incomes" (Myint, 1987, p.110) and to direct this investment surplus towards a broad base of manufacturing industries. The timetable of Rostow's project is also very precise. Rostow centres the development process upon a period of take-off - a catastrophist metaphor itself - which is threatened continually by a vicious circle of low per capita incomes, low rates of savings and rapid population growth. In these circumstances marginal adjustments in the mould of monoeconomics will not sponsor the structural transformation upon which development depends. Capital must instead be mobilised by the state and if necessary be supplemented by direct foreign investment and by official development assistance.

Rostow was not the only protagonist of a big-push. The term is more properly associated with Paul Rosenstein-Radan (1943) and with Tibor Scitovsky (1954). Both these authors insist that static market prices fail to communicate a range of external economies and diseconomies central to the development *process*. This being so, "to capture the alleged pecuniary external economies, of which the private producers are supposed to be unaware, simultaneous expansion of all industries is necessary. Only complete integration of all industries can do the job. This amounts to a plea for comprehensive central planning" (Haberler, 1987, p.79). It also describes a big push. According to Rosenstein-Radan (1943), the state must take responsibility for the simultaneous expansion of many industries even as it finances a vast investment in social overhead capital. Again, the capital is

to come both from abroad and from the directed savings of an indigenous rural sector. A second set of models is focused on the issue of urban-rural relations. A common starting point is the Prebisch-Singer thesis on the supposed secular deterioration in the terms of trade of weakly industrialised countries. This 'model' provides an obvious rationale for the urban-industrial bias which critics detect in orthodox development economics. In support of import-substitition industrialisation, the model legitimises both an increased local dependence upon food aid and an overvalued exchange-rate, the latter to spur an initial importation of machines and capital goods. The urban bias of Keynesian economics is said to be further inscribed in the dual economy/labour surplus models of Lewis (1955) and Nurske (1959). Together, these models embody a series of 'orthodox' propositions regarding the developmental function - or lack of it - of traditional agriculture. The most obvious assumptions are:
(i) that agricultural opportunities are the least attractive source of economic growth. Agriculture is beset by diminishing returns, while in industry economies of scale are the norm;
(ii) that agriculture can provide a substantial part of the capital that is required to mount industrialisation. Agriculture can provide an initially unlimited supply of labour for industry and it can provide much of this labour at zero opportunity cost. The reason for this is the heavy disguised unemployment to be found in rural areas of the Third World;
(iii) that policies and administrative means are required to keep food prices down in favour of urban consumers and to promote industrialisation. This is unlikely to dampen food output because farmers in developing countries are presumed to behave perversely. One implication of the labour surplus model is that the supply curve of farm production is backward sloping. Further, the indivisibilities of modern agricultural inputs are such that large farms - serviced by the industrial sector - are required to produce farm products at minimum costs (after Schultz, 1987, p.20).

A final set of models makes its seal with economic policy on the 'urban-industrial' side of the equation. In line with the big-push/labour surplus paradigm, orthodox development economics is said to promote sectoral and spatial planning regimes geared toward heavy industrialisation and big-city growth. The Keynesian legacy is here styled as a central-ising, top-down philosophy of development. Governments of developing countries are encouraged to nationalise the commanding heights of their economies and then to filter the benefits of rapid industrial growth down a central-place hierarchy. The pursuit of 'balanced growth' in turn is fostered by the building up of a set of growth centres (see Gore, 1984) and by the co-ordination of the investment strategies of a group of large industrial complexes, (cf. the Mahalanobis strategy in India). From the perspective of its counter-revolutionary critics, development planning brings to life the main tenets of orthodox development economics. It shares the commitment of the latter not just to statism and to dirigisme, but also to centrism and to urban bias.

The neo-classical resurgence

A neo-classical critique of Keynesian development theory is implicit in its attempts at characterisation. We can return, first, to the intellectual context fashioned by the Depression and its aftermath. Counter-revolutionaries challenge the view that the Depression represented a crisis of competitive capitalism. According to Haberler (1987, p.56), "the theory of secular stagnation is a gross misinterpretation of the Great Depression. Unfortunately, it was taken up by Raul Prebisch and thus had a strong impact on development economics". For Haberler (1987, p.56), "the Depression of the 1930s would never have been so severe and lasted so long if the Federal Reserve had not by horrendous policy mistakes of omission and commission caused or permitted the basic money supply to contract by about 30%. One need not be an extreme monetarist to recognise that such a contraction of the money supply must have catastrophic consequences. In other words the Great Depression was not a 'crisis of capitalism', as Prebisch says, but was a crisis of largely anticapitalist government policy, the consequences of horrendous policy mistakes".

Haberler is not the only counter-revolutionary to reason thus. His suggestion that market failure is possible under capitalism, but that secular stagnation is not, is taken up by Lal (1983), by Schultz (1987) and by Little (1982). This coterie further suggests that market 'failures' are most often caused by excessive state interference: by protectionism, by exchange controls, by deficit financing and so on. The lesson for development studies then must be that market capitalism, properly run, will secure long-run growth and development in both North and South.

The last point returns us to the monoeconomics versus duoeconomics controversy. The logic of the neo-classical critique will now be obvious. Because unemployment and declining commodity prices are not the rule under capitalism, it follows that heavy disguised unemployment and declining terms of trade in the developing world are equally undemanding of a duoeconomics which assumes their existence to be fixed and eternal, save only for the fix of industrialisation. To return again to Haberler (1987, p.61-62), "I have been critical of the view underlying much of development economics that developing countries as a group are set apart from the developed countries and are disadvantaged; that they are characterised by heavy 'disguised' unemployment; that their terms of trade have an inexorable tendency to deteriorate .... that they are subject to pernicious demonstration effects; that private initiative and market forces can be assigned only a minor role; that development requires 'balanced growth' on a large scale and a 'big-push' brought about by comprehensive 'programming' by the government". On the contrary, seen through counter-revolutionary eyes, "it is obvious that the developing countries are a very heterogeneous group" (Haberler, 1987, p.62). It is further evident that developing countries which pursue liberal, market-oriented policies, "are not bothered by the handicaps and afflictions .... from which all developing countries are supposed to suffer. Their success is fully explained by, and confirms, the neo-classical paradigm" (Haberler, 1987, p.62).

The demand for an 'outward-looking' development strategy feeds through into the debates on agriculture and urban bias. In the most recent Reports of the World Bank (those associated with Anne Krueger and most obviously the 1986 *World Development Report*), problems of poverty, of famine and of food production are placed squarely within the context of trade and pricing policies in world agriculture (World Bank, 1986, p.3-14 and p.61-84). The advice of the World Bank cuts two ways. On the one hand, industrial nations are warned against further imposition of trade barriers in agricultural markets. They are also criticised for putting in place agricultural support policies which combine to depress the traded prices of major foodgrains. On the other hand, governments of developing countries are blamed for pursuing "policies that inhibit the growth of agricultural output and of rural incomes" (World Bank, 1986, p.9). The Bank's target is a perverse system of global food production and exchange wherein, "most of the world's food exports are grown in industrial countries, where the costs of food production are high, and consumed in developing countries, where the costs are lower" (World Bank, 1986, p.8-9).

To break out of this vicious circle the World Bank urges "all countries [to seize an] opportunity to increase their incomes by specialising in products in which they have a comparative advantage" (World Bank, 1986, p.8-9). This is fully in line with counter-revolutionary thinking. In place of a too-expensive and too-optimistic big-push to industrialisation, developing countries are now counselled to find their niche in an expanding global division of labour. To the extent that this division of labour is market-driven, there is no reason to suppose, or fear, an unequal exchange of embodied labour time in manufactured and non-manufactured goods.

The counter-revolution's emphasis upon production for a global food market is complemented by its (re-)evaluation of the role and 'rationality' of farm households in the developing world. Once again there are subtle differences of tone and prescription depending on the authority one consults. In Eric Clayton's (1983) book, *Agriculture, Poverty and Freedom in Developing Countries*, a scholarly interest in farming systems research is lost amidst a diatribe against 'Leftists' and alleged fellow-travellers. For Clayton, smallholder agriculture in the Third World presents a model of enlightened self-interest and entrepreneurship. The demand for land reform - or "levelling down" (Clayton, 1983, p.43) - made by 'socialists and Marxists' is here judged an attack upon personal freedoms. Correspondingly, "the notion of 'relative' poverty [and basic needs] reflects a particularly unattractive attribute of Western man - that of envy .... [which] Leftists have exploited to increase discontent" (Clayton, 1983, p.29). For Clayton, variation in the size of peasant farms is more properly explained by "the variation of personal characteristics and qualities of small farmers and their families" (Clayton, 1983, p.37). Clayton's defence of smallholder agriculture and 'rural bias' gains no credence from statements such as these, or from his crude suggestion that, "Marxism, for obvious historical reasons, has nothing to say about agricultural development (particularly of peasant agriculture) and marxists have shown little understanding of the problems of farmers and farming" (Clayton, 1983, p.32). One must also wonder about the academic standing of a book whose author bemoans the opening up of "development discussion .... to sociologists" (Clayton, 1983, p.27). Says Clayton (1983, p.27), "This has resulted in a heavy influx of left-wing academics, and others, and increased the stream of development literature to a flood, much of it predictably of a sociological and left-wing nature".

Fortunately, a polemic against orthodox development economics can be run alongside a reasoned defence of 'traditional agriculture'. The work of Nobel laureate Theodore Schultz provides a case in point. With the publication of *Transforming Traditional Agriculture*, Schultz (1964) added spark to the neo-classical resurgence in development studies. Three main points emerge from Schultz's work and they remain central to the counter-revolution of the 1980s. Schultz first reclaimed the farmers of the Third World as productive and rational economic actors. Setting his stall against received notions of backward-sloping supply curves and disguised unemployment, Schultz (1964, p.35) argued that, "there are comparatively few significant inefficiencies in the allocation of factors of production in traditional agriculture". More recently, he has written: "Farmers the world over, when dealing with costs, returns and risks are calculating economic agents. Within their small, individual, allocative domain they are fine-tuning entrepreneurs, tuning so subtly that economists from high-income countries fail to see how efficient they are" (Schultz, 1987, p.31).

Schultz next considered how traditional agriculture might be transformed. Having adduced the allocative efficiency of peasant agriculture, Schultz dismisses the suggestion that further investment should be made in the existing stock of factors of production. He says this will be a costly strategy, given the low rate of return at the margin. It follows that traditional agriculture must be transformed from without. In terms reminiscent of Rostow, Schultz commends the 'jump starting' of Third World agriculture by means of the introduction of new inputs and technologies. Nitrogenous fertilisers and modern irrigation schemes comprise two such inputs, but Schultz's chief recommendation is for the adoption of the new high-yielding varieties of grain which partly comprise a Green Revolution. More recently, the work of Schultz has been extended by Hayami and Ruttan (1971) with their advocacy of a process of induced innovation in agriculture.

The concept of induced innovation in turn is linked to Schultz's third legacy to the counter-revolution; his insistence upon 'getting prices right' (see also Bauer, 1947). Says Schultz (1987, p.25), "In the case of poor performance of agriculture, the real culprit is the lack of economic opportunities that are rewarding to farmers". Specifically, Schultz blames the urban bias of Keynesian development economics and the governments in tow to this paradigm. In pursuit of an elusive import-substitution industrialisation the agricultural sectors of the Third World have been bled dry. On the one hand, the price of foodgrains is often depressed by government policy. Unsurprisingly, food production is then discouraged and, in the case of sub-Saharan Africa, per capita food production even begins to fall. Conversely, where food prices are allowed to approach a market rate, as in Deng's China, foodgrain production is massively stimulated (see Cheung, 1986). On the other hand, much needed agricultural inputs must either be imported or bought at the inflated prices charged by locally protected, import-substituting enterprises.

Put together, the criticisms advanced by neo-classical 'monoeconomists' tell tale of a gloomy development malaise. For such critics, the legacy of orthodox duoeconomics is written deeply and malignly into the landscapes of most developing countries. It is seen first in the hollow faces of the famine-affected populations of Africa. Says Schultz (1987, p.24), "The food tragedy in Ethiopia at present is in no small measure a consequence of the economic policy of that government". It is seen, second, in the low output performances of several 'agricultural nations' and in the underutilisation and undermechanisation of many rural landscapes. Again, "The poor performance of agriculture in the USSR, Poland and a fairly long list of other countries is not the fault of *nature*" (Schultz, 1987, p.25; emphasis in the original). It is seen, third, in the overburdened transportation systems which bring many able rural dwellers to the towns and cities of the Third World. It is seen, fourth, in the cities themselves, where urbanisation runs ahead of industrialisation and where recent migrants are condemned to live in slums and shanties and to eke out a living in the informal sector. [Note: The neo-classical discourse is ambivalent on the informal sector which, at one level, is praised for its risk-taking and dynamism. In the present context what matters is that protectionism and state-aided heavy industrialisation have helped create an overburdened informal sector.] It is seen, finally, in the 'empty' shops and long queues for restricted imported goods. In each case the heavy hand of the state is revealed in a torpid urban and regional landscape created directly in its image.

The counter-revolution considered

The counter-revolution's critique of duoeconomics implies a definite political agenda of its own. John Toye (1987, p.49) suggests that, "the promise of the counter-revolution in development policy is to solve the problem of inadequate development by neutralising three particular causes": an over-extended public sector wherein governments take on "functions which are beyond the proper, appropriate or normal functions of government, especially in the sphere of production"; an over-emphasis upon physical, as opposed to human, capital formation; and a proliferation of distorting economic controls. Obviously, different authors within the neo-classical pantheon assign different weights to each of these problems. (Authors also vary in terms of the rigour of their proposed solutions: compare the careful analysis of sub-Saharan agriculture made in the 'Berg Report' (World Bank, 1981) with Peter Bauer's (1984) apocalyptic denunciations of income redistribution as morally illegitimate, of most fixed capital formation and of the post-*laissez faire* state.) To Toye's list we might add a fourth problem in search of a solution: that of import-substitution industrialisation (ISI). ISI is to make way for an outward-looking, or export-oriented, development strategy. If the bulk of these exports are in the form of primary commodities, including foodstuffs, so be it; the hard currencies then earned can be used to buy in foreign grains and/or to strengthen the private capital pools available to farmers. In any case, once the distorting price systems associated with ISI have been removed, it is expected that local agricultural production will blossom and expand.

What are we to make of this 'new' prospectus? In answering this question it is necessary to highlight a tension at the heart of the counter-revolutionary discourse. The tension concerns the supposed independence of the economy (and of market economics) from the polity (and the political market). In arguing for a monoeconomic approach to the dilemmas of development, counter-revolutionaries strongly oppose statism and dirigisme, by which they understand a mistaken attempt by governments to run public industries and to regulate mechanisms of exchange and price formation. The state, in this discourse, is to be responsible only for the maintenance of law and order, for the control of the money supply, for the provision of basic health and educational services and for the establishment of basic communications (Bauer, 1959, p.96 and Toye, 1987, p.55). Several questions then arise. On what basis is this selection of tasks made and how is it to be defended? Can the separation of political and economic activities be made other than on political grounds? (John Toye makes an excellent point here. He notes that "it is precisely on political grounds that Bauer proscribes certain kinds of activity by the government .... The extent of the government sector in India was to be determined, according to Bauer, by the requirements of developing 'a society resistant to the appeal of a totalitarian regime', which was 'the essential American interest in India'. If that is not a political criterion for deciding whether or not the public sector is over-extended it is difficult to imagine one. Yet Bauer himself complains bitterly that development economists subordinate knowledge to public purpose" [Toye, 1987, p.57, quoting Bauer, 1959, p.95]). What political assumptions are written into the assumed independence of an equilibrating and fair 'market mechanism'? To condemn, as Bauer does, all forms of government investment for necessarily overriding the preferences of ordinary people, is to make a series of assumptions about the constitution of the economy, the state and civil society. The *laissez faire* position, which is not accepted by all counter-revolutionaries, assumes that the distribution of income and economic power within a society is, in some sense, given and natural; it is a product of the intrinsic and differential capacities of its economic participants. This allows Bauer and Clayton to inveigh against the 'immorality' of land reforms and other 'imposed' income redistributions. It also assumes the 'morality' of inter-generationally transferred income, wealth and power differentials. The neo-classical discourse further assumes that ordinary people are 'free to choose' and that the sum total of their freely exercised subjective preferences will secure an outcome that is efficient and equitable from the standpoint of a wider collectivity. This in turn implies that there is no 'isolation paradox' (to borrow a term from Amartya Sen); that there is no case "for government action to overcome the sub-optimality of savings and investment" (Toye, 1987, p.58), and that, despite market imperfections, there is no case for government actions as per the theorem of the second best. Finally, most counter-revolutionaries dismiss the case for rapid, exogenously impelled, 'development'. They assume the capacity of economic gradualism to perform this task, and to perform it as guided by a study of comparative statics. This last assumption surfaces most clearly in the renewed defense of comparative advantage theories.

To make explicit these economic and political assumptions is not to dismiss the counter-revolution in all its aspects. In the work of Haberler (1987), Schultz (1987), Little (1982) and Beenstock (1983), more so than in the polemics of Lal (1983), Clayton (1983), Johnson (1971) and Bauer (1981), one finds many thoughtful prescriptions for the 'development process'. In the case of international trade, Haberler (1987, p.73) is right to maintain "that the static nature of trade theory does not deprive it of usefulness in explaining dynamic processes .... static gains from trade along the lines of comparative cost enable a country to save and invest more. Furthermore, it attracts capital from abroad and fosters the importation of technical know-how. This means that the static production possibility curve is pushed out". This has some insistent implications for the reorientation of Third World agricultural systems (Killick, 1984). With regard to urban bias, also, the neo-classical critique is, in part, well-founded. The crude distinctions between a generative urban-industrial sector and a stagnating countryside, to be found in some early texts, did great harm when translated into development plans and policies. The food crisis in sub-Saharan Africa is intricately bound up with this 'misguided' industrialism, and with much else besides. Finally, the counter-revolution has added to our understanding of the role of markets, and the 'market mechanism', in the Third World. We can no longer assume that farmers and business people in developing countries are of necessity - and for cultural/psychological reasons at that - perverse in their reading of market signals and local systems of incentives and disincentives. Nor is there always much to be gained from detaching national systems of prices and exchange-rates further from the target range suggested by local factor endowments and global comparative costs. This is one lesson which has emerged from East Asia. Where market prices are formed according to dispersed power principles, and where they reflect local factor endowments, a sensitivity to market signals can secure a favourable integration into a new international division of labour.

In one sense, then, the counter-revolution has much to commend it, and it should not be dismissed simply because it makes a number of assumptions. All forms of discourse make assumptions. What is misleading about the counter-revolution, and what makes it potentially dangerous, is its attempt to conceal these assumptions. Neo-classical economics fails properly to examine the conditions under which its assumptions could, and will, be realised in developing nations. To see that this is so, consider just three aspects of the counter-revolution's 'hidden agenda': its caricature of orthodox and Marxian development economics;

its dismissal of development planning; and its attempts to depoliticise its own political ideology.

On orthodox and Marxian economics we must be brief. The reader may have noted already that the neo-classical 'resurgence' too often substitutes polemic for critique. To suggest, as Clayton (1983) does, that Marxists are by definition ignorant of peasants and the rural sector is crass and unjustified. Writers who take seriously the legacy of Marx have long inveighed against urban bias and against state industrial capitalism (see Davey, 1975; Evans, 1979; from an 'orthodox' perspective see Lipton, 1977; Johnston and Kilby, 1975; Mellor, 1976). They do so, however, as part of a wider critique of dependent and associated capitalist development. As such they situate the politics of urban bias firmly within the context of local class systems which have themselves taken shape in the shadow of imperialism and colonial rule (see Bates, 1981). Writing of India, Pranab Bardhan (1984) explains the poor performance of the agricultural sector in terms of a crisis of public and private investment, and a corollary leaning to import-substitution. This crisis-cum-bias in turn is called forth by the constitution and dynamics of the post-colonial state in India. The state in India exists as an 'intermediate regime'. A coalition of "the industrial capitalist class, the rich farmers and the professionals in the public sector" (Bardhan, 1984, p.54) here combines to frustrate those economic reforms which would favour either economic liberalisation or a radical redistribution of assets.

The counter-revolutionary discourse is not helped either by its blunt dismissal of the case for development planning. To begin with, it fails to acknowledge the strong case that can be made for state intervention and co-ordination. The theorem of the second best, the existence of manifold market imperfections, the infant-industry argument, the repeated demonstration of a Prisoner's Dilemma, all these are wished away in many. (Not all accounts: Haberler (1987) and Lal (1983) thus deal at length with the infant-industry argument and with 'second-best economics').

The counter-revolution also fails to distinguish between the rhetoric of state planning and its actual effectivity. Sukhomoy Chakravarty (1987) raises just this point in his discussion of development planning in India. Chakravarty notes that the high-point - or nadir - of development planning in India is supposed to have been reached with the Second Five Year Plan, 1956-61. At this time the Mahalanobis faction is said to have 'captured' the government in New Delhi and through the Planning Commission to have inflicted upon India the heavy hand of aid-fed, urban biassed, planned heavy industrialisation. The inefficiencies there engendered are supposed to have crushed India's agricultural sector and to have set the scene for the famine of 1965-67. Yet there is little evidence for this fiction. Under Mahalanobis a complicated input-output matrix was created and a slight drift to import-substitution industrialisation was initiated. But the bark of the Planning Commission was very much worse than its bite. Precisely because of the weakness of the development state in India, the much bemoaned 'massive shift' of resources from agriculture to industry probably never took place. Moreover, on a year-by-year basis, the intellectual foundations of the Mahalanobis Plan, which was clearly informed by the Lewis-Nurske models, was toned down in favour of a politically negotiated pragmatism. Planning in India - the *bête-noire* of counter-revolutionary critics - never has been as hegemonic as its critics wish to suppose.

Finally, the counter-revolutionary dismissal of planning is in part gainsaid by the successes of planning. Without wishing to claim too much for planning, or for a dirigiste world order, it is worth noting that the heyday of planning and the Bretton Woods world order coincided with "a tremendous growth of the world economy and of world trade .... and the emergence of many new industrial centres in parts of the world, including the Third World and the communist bloc" (Haberler, 1987, p.74); and that, in India, the economy 'as a whole' grew at a rate of 3.5 per cent per annum from 1951 to 1965, with the agricultural sector recording an average rate of growth of 2.6 per cent per annum. At the same time, manufacturing industry's share of GNP increased from 10 per cent to 13.6 per cent (Datt and Sundharam, 1986). It might also be noted that an element of urban bias is implicit in the development process, planned or otherwise. If we assume that "a decline in the relative importance of agriculture [is part of] the process of economic development", it follows that agriculture must constitute "the principal source of savings or capital accumulation and labour supply for the rest of the economy .... The question is not whether the non-agricultural sector should draw on agriculture [but] .... at what pace, by what method, and in what time sequence such intersectoral transfers of resources .... should take place" (Islam, 1987, p.39-40). To put it another way, the existence of urban bias should not surprise or alarm us. What matters is the scale, bases (class, region, gender, ethnicity) and direction of the transfers thus instated and the uses to which they are put.

This brings us to a third aspect of the counter-revolution's hidden agenda: its attempt to depoliticise its own political intentions even as it refuses all other political economies. The counter-revolution advances two propositions here, each of which is questionable. On the one hand, the alleged failures of ISI and of slow agricultural growth are laid at the door of a misguided and rent-seeking political elite, or set of elites. (If anything, it is the hegemony of 'ideas' that receives the sharpest criticism.) On the other hand, counter-revolutionaries will claim the 'success stories' of East Asia - and the NICs especially - as examplars of a purely economic, because, market logic. In Haberler's (1987, p.62) words: "Their success is fully explained by, and confirms, the neo-classical paradigm". Put together, these propositions allow the counter-revolutionaries to promote a powerful ideology of economic liberalism. Where the state eschews 'politics' for the hidden hand, a wave of economic development can be expected to break swiftly over the spatial and social landscapes of a developing nation.

Two examples will expose the dangers of such an assertion. Consider, first, the so-called Green Revolution in South Asia. Writing of the Indian Punjab, Theodore Schultz claims that: "correct governmental actions do occur" (Schultz, 1987, p.33) - his case in point being the decision of India's Ministry of Agriculture to import new Mexican dwarf wheat seeds in 1966; and that the "success-story" that is Punjab today could have been replicated throughout India had not the Government - fearing adverse social consequences - removed the structure of incentives which so aided the "farm entrepreneurs of the Punjab" (Schultz, 1987, p.33).

Doubtless, there is something in each of these propositions. What Schultz fails to discuss, however, is at least as important, and for this reason most instructive. To take a minor point first: Schultz fails to tell us why the dirigist state in India did take at least one correct development decision. More worryingly, Schultz fails to consider the political economy (and social context) of 'producer incentives' in Indian agriculture. One reason why the government was warned of adverse social consequences was precisely that conditions of access to Green Revolution technologies in India are not equal. Many farmers lack the capital or credit necessary to purchase tubewells and pumpsets, fertilisers and pesticides. Nor is it the case that non-adopting farmers and labourers must gain from an increased demand for labour associated with the Green Revolution. The question of employment and wages will be settled according to local labour market conditions and capacities for capital substitution. Further, these labour markets are themselves constituted unequally; power is not dispersed in the atomistic and diffuse patterns demanded by neo-classical economics. The *jajmani* system still holds sway in large areas of rural India, alongside a series of more or less coercive (or bonded) labour markets. Finally, it is misleading to ascribe the 'success' of the Green Revolution in Punjab entirely, or even mainly, to the accident of favourable producer prices for foodgrains. Still more important were "a well-developed irrigation system, a long-standing system of land-owning cultivators, adequate rural infrastructure, including roads, markets and rural electrification, and a seed and fertiliser technology supported by national agricultural research" (Islam, 1987, p.43). More generally, one would wish to emphasise the commitment to the Green Revolution of the United States and the development state in India. One 'paradox' which seems lost upon the critics of urban bias is the extraordinary extent to which the state in India has subsidised and propagated a 'peaceful' transformation of its countryside (see Rudolph and Rudolph, 1987). At the regional, or State, level especially, the state in India is dominated either by landlords and bullock capitalists and/or by merchant capital (Harriss, 1984).

A second example concerns international trade and the new international division of labour. Neo-classical theorists continue to laud the achievements of East Asia's NICs and China today, in negotiating a favourable position in a new global market-place. Apparently, this is because they have got their prices and policies right (Balassa, 1982). Yet 'orthodox' theorists have demonstrated quite as often that the preconditions for take-off in each of these cases were met in large part by an interventionist state. It was the state in Taiwan, and in South Korea, which redistributed economic assets on a more egalitarian basis, and which actively subsidised export-substituting industries until these industries had a comparative advantage which corresponded to local factor endowments (see Hamilton, 1987). To extend this logic to parts of Africa and South Asia could be disastrous. In the absence of prior structural reforms a concerted push to serve 'the market' might further detach local trajectories of growth from development. To paraphrase Bhagwati, 'growth could be immiserating'. Moreover, if 'successful', such a policy would surely run foul of protectionism in the industrial nations. It is a fallacy of composition to suppose that all developing countries can run a balance of trade surplus at the same time.

A not dissimilar argument may hold in the case of the agricultural sector. It is not unreasonable for the World Bank and others to urge various African and Asian nations to export foodgrains and other commodities on the world market, and then to use their hard currencies to make good a possible food gap within a country. This is one way of meeting basic needs. Nevertheless, it is vital that certain preconditions of this strategy are first met, and that the assumptions of a 'global ranch-cum-supermarket' are fully understood. In this case one might ask why certain developing countries are being asked to forsake ISI for a commodity export-based development strategy. Might this have something to do with the crises of global Fordism (Lipietz, 1987) and with the industrial countries' own fight for industrial export markets? Further, what do we know of the composition of international food and commodity markets? In what sense is there a global market in foodgrains when that market is dominated by just five 'merchants of grain'? What evidence is there of the price elasticity of exported primary commodities? Is it not the case that for many commodities demand is very price-inelastic, and that such demand, geographically, is dominated by importers in a few, powerful, nation-states? Finally, what reason is there to suppose that the substitution of export crops for local foodgrains will not result in a crisis of nutrition in some areas of the Third World? In the real world of imperfect and dominated markets, can we be certain that the direct entitlements of poor families to food will be fully replaced by market or state-directed transfers? The answer must be 'no'.

Conclusion

The failure of neo-classical economics to address these and other questions must cast a shadow over its supposed relevance to the Third World. Far too often, the counter-revolution proceeds by tilting at windmills. Judged against the worst excesses of import-substitution industrialisation and urban bias, an appeal to the market can seem attractive. But how many 'orthodox' (or Marxian) economists today subscribe to the Lewis and Nurske models or to Rostow's typology of modernistion? Very few, one suspects (and still fewer students outside of the 'dismal science'). Moreover, the neo-classical theorist can only account for the failures of ISI and planned development by pointing up the 'flawed logic' of duoeconomics. The economics of 'distortion' and rent-seeking are here blamed upon the poverty of development *theory* and upon those 'bureaucratic' systems which encourage rational actors to behave corruptly. What the counter-revolution quite fails to discuss is the wider background of dependent and associated development against which these systems and structures are (variously) produced and reproduced. Neo-classical economics proposes an essentialist discourse in which 'the laws of economics' are derived, *tout court*, from a series of assumptions about human nature: most obviously, that individual economic actors maximise their utilities and are endowed with different tastes (and therefore wants) and productive capabilities (see Wolff and Resnick, 1987).

To argue thus is not to tilt the balance back towards the state in some abstract debate about state versus market. The point is that there are *states* and *markets*. What is important is the constitution and interaction of these varied sites of resource appropriation and distribution; at issue are the relations of power and conditions of access that are built into these sites and which are continually contested, reshaped and re-built. What matters, too, is that the counter-revolution and international capital are not given free rein to promote an abstract 'market mechanism' as the sole producer of space and nature in developing countries. In the context of urban-rural relations there surely is a need to contest the logic of 'urban bias'. No doubt, too, a restoration of incentives in agriculture will serve the cause both of equity and efficiency, and in the process help stem the tide of rural-urban migration. But the market's mythical powers need to be treated with caution and some mistrust. Amidst the current crises of global Fordism, there is reason to beware a 'market principle' which threatens to cut short industrial deepening in the Third World and which threatens to sacrifice local entitlements to local foodgrains. Under sanction from the World Bank and the IMF there is reason to fear that some developing countries will be reallocated their 'nineteenth century' niche in the international division of labour (Corbridge, 1988). Finally, there is reason to expect that the new 'market realism' will be enforced, domestically, by the same propertied classes which have long controlled the 'development state' and which have long benefitted from ISI. Once again, the politics of intra-sectoral production and consumption may be brushed aside. The difference this time is that it will be done in the name of *the market* and not of *the state*. Vive la difference.

REFERENCES

Balassa, B. and Associates (1982) *Development Strategies in Semi-Industrial Countries*, Johns Hopkins University Press, Baltimore

Balassa, B. (1985) 'The Cambridge Group and the developing countries', *The World Economy* 8, 201-18

Bardhan, P. (1984) *The Political Economy of Development in India*, Blackwell, Oxford

Bates, R. (1981) *States and Markets in Tropical Africa*, University of California Press, Berkeley

Bauer, P. (1947) 'Malayan rubber policies', *Economica*, May 14, 179-87

Bauer, P. (1959) *United States Aid and Indian Economic Development*, American Enterprise Association, Washington DC

Bauer, P. (1981) *Equality, the Third World and Economic Delusion*, Methuen, London

Bauer, P. (1984) *Reality and Rhetoric: Studies in the Economics of Development*, Weidenfeld and Nicolson, London

Beenstock, M. (1983) *The World Economy in Transition*, Allen and Unwin, London

Chakravarty, S. (1987) *Development Planning: The Indian Experience*, Clarendon, Oxford

Cheung, S. (1986) *Will China "Go Capitalist"?*, Institute of Economic Affairs, London

Clayton, E. (1983) *Agriculture, Poverty and Freedom in Developing Countries*, Macmillan, London

Corbridge, S. (1986) *Capitalist World Development: a Critique of Radical Development Geography*, Macmillan, London

Corbridge, S. (1988) 'The Third World in global context', in: Pacione, M. (ed.) *The Geography of the Third World: Progress and Prospect*, Routledge, London, 29-76

Datt, R. and Sundharam, K. (1986) *Indian Economy*, Chand, New Delhi

Davey, B. (1975) *The Economic Development of India: a Marxist Analysis*, Spokesman, Nottingham

Evans, P. (1979) *Dependent Development: the Alliance of Multinational State and Local Capital in Brazil*, Princeton University Press, Princeton

Forbes, D. (1984) *The Geography of Underdevelopment*, Croom Helm, London

Gore, C. (1984) *Regions in Question*, Methuen, London

Haberler, A. (1987) 'Liberal and illiberal development policy', in: Meier, G. (ed.) *Pioneers in Development, Second Series*, Oxford University Press/IBRD Oxford, 51-83

Hamilton, C. (1987) 'Can the rest of Asia emulate the NICs?', *Third World Quarterly* 9, 1225-56

Harriss, B. (1984) *State and Market*, Concept, New Delhi

Hayami, Y. and Ruttan, V. (1971) *Agricultural Development: an International Perspective*, Johns Hopkins University Press, Baltimore

Hirschman, A. (1982) 'The rise and decline of development economics', in: Gersovitz et al. (eds.) *The Theory and Experience of Economic Development: Essays in Honour of Sir W. Arthur Lewis*, Allen and Unwin, London

Islam, N. (1987) 'Comment on Schultz', in: Meier, G. (ed.) *Pioneers in Development, Second Series*, Oxford University Press/IBRD Oxford, 39-48

Johnson, H. (1971) 'A word to the Third World: a western economist's frank advice', *Encounter* 37, 3-10

Johnston, B. and Kilby, P. (1975) *Agriculture and Structural Transformation*, Oxford University Press, Oxford

Killick, T. (1984) 'Economic environment and agricultural development: the importance of macroeconomic policy', *Food Policy* 10, 29-40

Krueger, A. (1978) *Liberalization Attempts and Consequence*, Ballinger, Cambridge, Massachusetts

Lal, D. (1983) *The Poverty of "Development Economics"*, Institute of Economic Affairs, London

Lewis, W. (1955) *The Theory of Economic Growth*, George Allen and Unwin, London

Lipietz, A. (1987) *Mirages and Miracles: the Crises of Global Fordism*, Verso, London

Lipton, M. (1977) *Why Poor People Stay Poor*, Temple Smith, London

Little, I. (1982) *Economic Development: Theory, Policies and International Relations*, Basic Books, New York

Mellor, J. (1976) *The New Economics of Growth*, Cornell University Press, Ithaca

Myint, H. (1987) 'Neoclassical development analysis: its strengths and limitations', in: Meier, G. (ed.) *Pioneers in Development, Second Series*, Oxford University Press/IBRD, Oxford, 107-36

Nurske, R. (1959) *Patterns of Trade and Development*, Almqvist and Wiksell, Stockholm

Peet, R. (ed.) (1987) *International Capitalism and Industrial Restructuring: a Critical Geography*, Allen and Unwin, Boston

Prebisch, R. (1961) 'The economic development of Latin America and its principal problems', *Economic Bulletin for Latin America* 7, 1-22

Reading, P. (1986) *STET*, Secker and Warburg, London

Rosenstein-Radan, P. (1943) 'Problems of industrialization in east and south-eastern Europe', *Economic Journal* 53, 202-11

Rostow, W. (1960) *The Stages of Economic Growth: a Non-Communist Manifesto*, Cambridge University Press, London

Rudolph, L. and Rudolph, S. (1987) *In Pursuit of Lakshmi*, Chicago University Press, Chicago

Schultz, T. (1964) *Transforming Traditional Agriculture*, Yale University Press, New Haven

Schultz, T. (1987) 'Tensions between economics and politics in dealing with agriculture', in: Meier, G. (ed.) *Pioneers in Development, Second Series*, Oxford University Press/IBRD, Oxford, 17-38

Scitovsky, T. (1954) 'Two concepts of external economics', *Journal of Political Economy* 62, 143-51

Smith, N. (1984) *Uneven Development*, Blackwell, Oxford

Toye, J. (1987) *Dilemmas of Development: reflections on the Counter-Revolution in Development Theory and Policy*, Blackwell, Oxford

Wolff, R. and Resnick, S. (1987) *Economics: Marxian Versus Neo-Classical*, Johns Hopkins University Press, Baltimore

World Bank (1981) *Accelerated Development in Sub-Saharan Africa: an Agenda for Action*, Oxford University Press/IBRD, Oxford

World Bank (1986) *World Development Report 1986*, Oxford University Press/IBRD, Oxford

9

RURAL-URBAN INTERACTION IN BARBADOS AND THE SOUTHERN CARIBBEAN: PATTERNS AND PROCESSES OF DEPENDENT DEVELOPMENT IN SMALL COUNTRIES

ROBERT B. POTTER

Introduction

Conceptualising 'development' in the Caribbean region from the point of view of an urban-rural dichotomy is highly unrealistic when either the history or present-day realities of the territories making-up the region are considered. The fact that the adoption of an urban-rural divide should be so unproductive reflects a large number of closely related factors, some of which it can be argued are generic in the sense of applying to all Third World nations, whilst others reflect the particularities of the Caribbean region itself.

The principal aim of the present chapter is to examine the factors which point to the counterproductive nature of regarding rural-urban relations as giving rise to a schism or clear divide. In so doing, the implications that these arguments carry both for geographical research based in developing countries and also in connection with the effective implementation of development programmes and plans within them, will be touched upon. In this respect, it will be argued that although the physical structure, limited resource bases, relatively simple economies, and in particular, extreme small size of Caribbean territories all serve as important local factors requiring the adoption of a spatially and socially integrated view of the development and planning processes in the region, the primary aetiological factor is that the whole socio-cultural history of the Caribbean has been inextricably bound-up with the process of colonial development from the fifteenth century onwards. It will be shown how these developments were effective in setting up complex patterns of interaction between different parts of the national territory, and that these spatial configurations have become fossilised in socio-economic space, so that they continue to mould and dominate present-day economic and social structure.

In the Caribbean, the local variant of dependency is the *Plantation Economy* which has come to be associated with the syndrome of what the New World group of social scientists have, following George Beckford, described as *Persistent Poverty* (Beckford, 1972). It is here that the argument presented in the present chapter dovetails with that of the volume taken as a whole, for it may be asserted strongly that the manifest and widely apparent *differences* which exist between urban and rural areas in most Caribbean countries are the outcome and reflection of a *unified* set of socio-economic and cultural processes that we customarily refer to as 'development', or socio-economic change (Brookfield, 1975). The operation of this unified and holistic process of change is attested by the existence of marked linkages and flows between what may all too readily be discerned in a simplistic analysis as quite separate rural and urban zones. These differences relate to movements and flows of a variety of types and magnitudes, for example, those of agricultural produce, people, factors of production, capital, ideas, innovations and the like (O'Connor, 1983). In the space available in the present account it is necessary to take a limited perspective, and latterly empirical attention is therefore focussed principally upon the clearly apparent differences existing between rural and urban areas in the provision of key retail, commercial, health and other facilities, and the ways in which these both reflect past, and continue to promote future, strongly marked symbiotic-parasitic behavioural relationships between such areas. In this connection, Barbados is used as a specific example. But prior to this, a more general conceptual-cum-theoretical view is provided of urban-rural relations and contrasts in the wider Caribbean region, both historically and with regard to the present-day.

Urbanisation, Development and Territorial Relations in the Caribbean: an historical view

The parallel pointed to previously, between strikingly apparent socio-economic differences on the ground on the one hand, and strong functional linkages and spatial interactions between such areas on the other, is a feature that needs specific elbaoration in the context of the Caribbean region. In this section, this issue is addressed, both in relation to theory and practice, starting with the latter.

In most parts of the developing world, in the period since the Second World War, urbanisation has come to be an important, albeit a problematic process (Mountjoy, 1976, 1978, 1982a). This is certainly true of the Caribbean, as recently discussed in a number of publications (see Clarke, 1974, 1975; Cross, 1979; Hope, 1983, 1986; Potter, 1985a). At the present time, around 52.2 per cent of the total population of the region is to be found living in areas which may be classified as urban. Often, the urban population is increasing at rates far in excess of those recorded by the rural fraction. Thus, annual rates of urban population increase of between 3 and 6 per cent are typical, and as Clarke (1974) has noted, rates twice as high as these are quite common. For the Caribbean region taken as a whole, United Nations' projections indicate that during the decade 1980-90, whilst the urban population will have been increasing at an annual average rate of 3.18 per cent, the comparable rural growth rate will have stood at only 0.71 per cent, falling below that of the urban sector of the population by a multiple of approximately 4.5 times (United Nations, 1975; Hope, 1986). In fact, the urban proportion has increased from levels of 38.2 per cent in 1960, and 45.1 per cent in 1970, and it is anticipated by the year 2000, will be somewhat in excess of two-thirds of the population, specifically standing at 64 per cent (United Nations, 1975; Hope, 1986).

In a recent overview of the urban process in the Caribbean region, the present author has demonstrated how the pattern of contemporary urbanisation stands with respect to the individual countries and groups of countries making up the region (Potter, 1985a, p.66-79). The map summarising these patterns is reproduced here as Figure 9.1,

Figure 9.1: Levels of urbanisation and urban primacy in the Caribbean region (Source: Potter, 1985a)

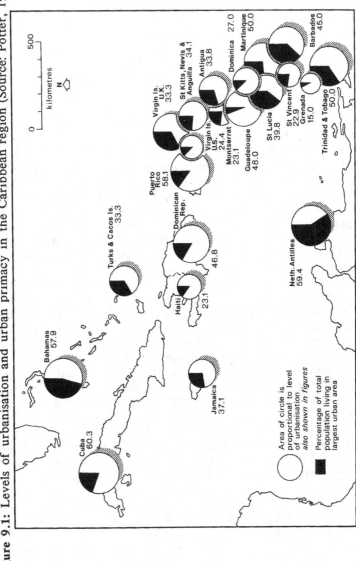

and serves to indicate - for the latest available date, generally in the middle to late 1970s - overall levels of urbanisation and urban primacy for countries within the region. The figure illustrates many of the well-recognised outcomes of the urban process in the developing realm. Firstly, a number of countries already show very high levels of urban development, among them Cuba (60.3 per cent), Netherlands Antilles (59.4 per cent), Puerto Rico (58.1 per cent), the Bahamas (57.9 per cent) and Trinidad and Tobago (50.0 per cent). The map also serves to summarise the extremely high levels of urban primacy which characterise these relatively small island economies. Often at least 30-40 per cent of the total population is to be found living in the single largest settlement. Further, Figure 9.1 invites the visually derived suggestion that levels of urban primacy are on average higher for the More Developed Countries (MDCs) in the region, good examples being provided by the Bahamas (c.46 per cent), Barbados (38.8 per cent), and Trinidad and Tobago (30.9 per cent). In fact, the analysis summarised in Potter (1985a, p.67-70) shows that in the Caribbean region, the correlation between levels of primacy and Gross Domestic Product per capita stands at +0.31. Further, a statistical relationship of +0.74 pertains between overall levels of urbanisation and G.D.P., whilst primacy is also demonstrated to be positively associated with overall levels of urbanisation at +0.43.

These summary observations agree with the thesis put forward by El-Shakhs (1972) that if Third World countries are considered as a group on thier own, a positive rather than a negative relationship frequently characterises the association between levels of urban primacy and economic development. This can be viewed alongside the fact that in virtually every Caribbean country, relatively strong urban primacy is the norm (Clarke, 1974; Potter, 1985a, 1985b), and this is generally associated spatially with a remarkably linear-coastal concentration of population and economic activities which may be likened to the settlement pattern envisaged in Vance's (1970) *mercantile model*, as an historically more fitting portrayal of the central place process.

This type of settlement evolution has recently been discussed with respect to the Caribbean region by the present author (Potter, 1985a, 1985b). The first stage of the Vance sequence involves the initial search phase of mercantilism, where the prospective colonising power seeks to collect information about economic opportunities and prospects as the precursor to the second stage, wherein any natural storage that the territory offers by the way of timber, furs, skins or fish is harvested. These two stages can, of course, be equated broadly with Christopher Columbus' initial chance discovery of the West Indies in 1492, and his 3 subsequent voyages of exploration between that date and 1504. In the Vance sequence, only as a third phase are permanent settlers established, these effectively forming the initial colony. In the Caribbean, this occurred at different times on various islands. However, in all cases, the pattern was similar, with settlers clearing the land of its natural vegetation and proceeding to plant and grow staple agricultural products. From this time onward, the colony starts to consume increasing volumes of manufactures produced by the mother country, and in return, supplies it with agricultural produce. By this juncture, symbiotic economic circumstances come to be reflected in the settlement system of the colony, which focuses almost exclusively on one coastal point of attachment. The broad sweep of this process has been elaborated for one Caribbean country, Barbados, in a previous publication (Potter, 1985a, p.70-72). Hence, as Clarke (1974) has summarised, whilst Caribbean urban settlements were administrative and commercial centres, they were never the locus for manufacturing activities of any salience. From the time of their first creation, Caribbean towns and cities acted as the intermediaries in a predominantly unidirectional set of linkages and interactions which could be traced from the internal rural staple producing areas to the external metropoles of Europe. The full implications of these highly skewed, asymmetrical and uneven sets of interactions will be elaborated subsequently.

The main period of colonisation came in the sixteenth and seventeenth centuries (Blume, 1974). During this period, the islands were colonised by the Spanish, Dutch, French, Danish and British, a process which can be dovetailed in broad terms at least, with the final two of Vance's five stages of settlement evolution. During this period of colonisation by

European powers, the key role must be ascribed to the plantation. Beckford (1972) has provided a detailed and most penetrating review of the plantation system as an agent of change in the Caribbean region, arguing that it has been the dominant economic, social and political institution in the past, remains through to the present, and from all indications, will continue to be so in the future. Thus, any understanding of urban-rural interaction within the region must start from a consideration of the plantation system.

At the outset of his thesis, Beckford argues categorically that the plantation "was an instrument of political colonization", so that, "in short, it has fashioned the whole environment which the people of these countries have inherited" (Beckford, 1972, p.3). But whilst political colonization came to an end in the immediate post Second World War period, the plantation as an instrument of colonial control and domination has survived, so that "though the winds (of change) may have swept, they have not swept clean" (Beckford, 1972, p.4).

In simple terms, a plantation may be defined as an agricultural unit which employs large numbers of unskilled labourers, who operate under a close system of supervision (Beckford, 1972), and which normally specialises in the production of only one or two marketable products. During the period of colonization, the Caribbean became a series of monocrop plantation economies, supplying staples primarily for export and external consumption in Europe. Whilst some might argue the economic efficacy of this in terms of the law of comparative cost advantage (Graham and Floering, 1985), others are inclined to see it as the start of a system of exploitation and expropriation at a global scale which has persisted unchanged to the present day.

Beckford (1972, p.14-15) presents evidence, showing that on a worldwide scale, the greatest concentration of longstanding plantation economies is to be found in the Caribbean. Thus, for example, both Antigua and Barbados were 'developed' as monocrop plantation economies producing sugar. In the latter case, this remained the mainstay of the economy for three hundred years. In similar fashion, Trinidad was principally dominated by sugar plantations, but coconuts and cocoa production was also involved. In some of the smaller islands of the southern

Caribbean, notably Dominica, St. Lucia, St. Vincent and Grenada among them, whilst the plantation system was somewhat less dominant, monoculture - in these cases related to banana production - and all of its vicissitudes, remained the economic reality.

The process of European colonisation witnessed virtually the complete decimation of the indigenous population in the Caribbean (Beckford, 1972). As the operation of plantations as production units was premised on large volumes of unskilled labour, and such work was far from popular among potential migrants from the mother country, such labour demands were eventually met by the transplantation of forced labour, mainly from West Africa, under a system of slavery (Lowenthal, 1972; Beckford, 1972), a process which created an entirely new social order.

At the same time, most plantation economies co-existed from the earliest times with peasant farmers, who were often involved in producing exactly the same cash crops. But competition between these archetypal representatives of large- and small-scale agriculture was obviously far from equal. This was especially true when it came to land, both regarding access to it and its quality. Beckford cites the West Indies as the outstanding example of peasant-planter conflicts. On the abolition of slavery in the nineteenth century, the freedmen endeavoured to move away from the plantations in order to establish their own independent existences. But on several islands, particularly those where sugar monocropping was in full swing, the land had already been alienated by the plantations long before emancipation. Prime examples of this are afforded by Antigua, St. Kitts and Barbados. In yet other instances, although land was still to be found, it was generally either inaccessible, or of very low quality. Often the only land available for freed slaves was mountainous in the extreme. This was certainly true with respect to much of the Windwards, Trinidad, Guyana and Jamaica. Thus, whatever the local economic and environmental circumstances, peasant farmers in the Caribbean had, through time, to compete directly with the plantations for both land and labour, obviously on a grossly unequal basis. Such was the stranglehold held by the plantations that frequently, such small 'independent' farmers were forced economically to undertake wage labour on them.

Many commentators take the view, therefore, that while emancipation may in theory have witnessed the gaining of freedom, in practice, it merely witnessed maintenance of exactly the same patterns of economic and social domination within the Caribbean region as had existed under slavery (see Lowenthal, 1972; Clarke, 1986; Potter and Binns, 1988). Large-scale plantation-type enterprises have been able to retain their agricultural hegemony by virtue of their adoption of advanced techniques, the ease with which they can secure credit and capital, their advantage in carrying out research and a whole host of other factors.

In concluding his view of the role of plantations in the propagation of *persistent poverty*, George Beckford maintained that this all pervasive social, economic and political system had perhaps been most influential in giving rise to what he described as a "dependency syndrome in our psychological makeup" (Beckford, 1972, p.234). According to Beckford, the means of escaping its influence so that the majority of the people can be provided with a real stake in their country lies with changing the perceptions of individuals (on this general theme and its wider implications for planning and development in the Caribbean, see Potter, 1985a). For Beckford (1972, p.235) then, the "most intractable problem is the colonized condition of the minds of the people. Until we decolonize the mind, there is little hope that genuine independence can be achieved".

Although Beckford did not cite or refer directly to Andre Frank whose seminal work had been published some three years prior to the appearance of his own text, clearly he was referring to the individual psychological connotations of *dependency*. In fact, Frank's model and theory can be employed directly in order to analyse the implications of the development of plantation economies for urban-rural interaction in the Caribbean region (see also Girvan, 1973; Palmer, 1979; Clarke, 1986). When this has been considered in the historical setting, post Second World War and contemporary consequences of the perpetuation of the broad forces of domination and differentation will be reviewed briefly.

If the development of large tracts of the earth's surface has been dependent upon metropoles, then it holds that the evolution of towns in developing countries has also been part of that dependent process, specifically resting upon the accumulation of social surplus product - an excess of production over need - and the articulation of the capital so engendered (see Roberts, 1978; Harvey, 1973, 1985; Potter, 1985a). It was Frank, a Chicago trained economist who did much to promote a world political economy approach to development studies, stressing that the development process depends on the role that the nation plays in the international economic system. His ideas were outlined in *Capitalism and Underdevelopment in Latin America* (Frank, 1969), where he argued that the condition of Third World countries is not a result of a lack of economic success or a failure to modernise, but a direct reflection of the manner of their incorporation into the international capitalist system. Thus, originated the thesis of the 'development of underdevelopment', or what is now generally referred to as *dependency theory*.

Under such a formulation, development and underdevelopment are but the opposite faces of the same coin. If this is true internationally with respect to the development of countries, it will likewise hold for the constituent areas which make up individual territories and, thus what can be referred to as a "Russian-doll" approach to development emerges. Such an approach can help explain both the differences between so called rural and urban areas within nations, and the types of interaction and linkage which bond them together within the global hierarchical system of dependent relations. According to this view, the underdevelopment of one set of areas results from their being at the bottom of the international hierarchy of dependence. The basically rural underdeveloped regions, referred to by Frank as 'peripheries', are held back by their very involvement with the metropoles of the world economic system. In fact, Frank maintained that the more that underdeveloped countries and regions become associated with the metropoles, the more underdeveloped they are likely to be, and not the other way around. This is so, according to Frank, for their association with the capitalist free-market system means that the very internal transformation which is required for true development to occur is rendered impossible. Although there are notable exceptions to this relationship (Gilbert, 1985), saliently from the point of view of the present discussion, Frank specifically cited the West Indies, along with north-east Brazil as examples of the development of underdevelopment. At the present time, of course, it is difficult not to ponder the fact that it is two of the most 'developed' Third World countries, Brazil and Mexico, which display the most desperate international debt crises.

It can be argued that the processes outlined by Frank and other dependency theorists in explaining these contradictions of world development are the same agents that have come to dominate urban-rural relations in developing countries. The principal causal feature is the existence of a chain of exploitative relations which involves the extraction and spatial articulation of social surplus product, via a process of *unequal exchange*. Thus, the terms of trade are always set in favour of the next higher link in the chain of world economic domination. In the Caribbean context, the peasant farmer is disadvantaged with respect to the plantation farmer, but essentially both rural forms of production witness the ultimate extraction of social surplus value by higher agencies (see also Beckford, 1975). Thus, the world price for sugar has, for instance, always seen labourers toil long and hard to eke a living, whilst as a result, sugar has been a cheap and abundant commodity in Europe. Surplus is also extracted in other manners, for example, by merchants and other intermediaries, retailers, commercial enterprises and professionals, their mainly being located in the primate (market) town. Surpluses in the form of profits are then transmitted to the metropolitan centres of Europe, principally in the guise of investment and expenditure, on, for example, the education of planter's children and direct consumption.

From the sixteenth century onward, the expansion of capitalism can, according to this view, be seen as synonymous with colonialism and underdevelopment, so that Caribbean towns effectively became, as was noted previously, administrative and commercial nodes for the collection and transhipment of social surplus product. In this way, whilst so called urban and rural areas are sharply etched, the contrast between them is the outcome of a single process. A reminder may be included at this juncture concerning the appropriateness of Vance's mercantile model of settlement and economic structuring, reviewed earlier in the chapter. It may be recalled that this is premised on the operation of 'exogenic' or outside forces, as opposed to the 'endogenic' or internal ones involved in Christaller's classical theory of the evolution of spatial economic structuring.

It is not then overstating the case to talk of the political economy of rural-urban relations and interaction in the Caribbean region, both historically and at the present time. Even without secondary industries which might have demanded a sharply polarised pattern of development, tertiary services and the outward signs of consumption became the hallmarks of concentrated and skewed urban development in Caribbean territories. Such levels of polarisation of infrastructure, social facilities and activities were predicated upon external demand and social influences, and not the needs and aspirations of the indigenous population. It is difficult not to reach the conclusion that the mercantile-capitalist periods led to a far greater level of spatial concentration than was ever socially desirable within what were after all very small non-industrial economies. Simple location theories tend to predict that without state intervention on the basis of social equity, pure economic forces of risk minimisation in achieving satisfactory profit levels will tend to promote the excessive spatial concentration of tertiary and quaternary services (see for example, Lösch, 1940; von Boventer, 1962; Webber, 1972; Clapp and Richardson, 1984, for instance). As a result, strong and enduring centripetal forces were set in motion within Caribbean countries, with rural folk travelling to the town for essential provisions, goods and services, to visit government departments and ministries, to collect wages and to complete all manner of what are really quite routine tasks.

Thus, as argued throughout this account, urban and rural have always been seemingly sharply contrasted portions of the national territory, but at the same time, ones which are very strongly linked by the chain of dependency. This is concisely summarised in the case of Barbados, which is on occasion described as "a city where sugar cane grows in the suburbs". This description serves to pinpoint the difficulty that is experienced if one tries to view social, economic and environmental conditions in Barbados and other Caribbean countries in terms of mutually exclusive rural and urban categories.

Indeed, on the ground in Barbados, the historical dominance of the plantation system is such that there are many more parallels between urban and rural areas than one would expect on the basis of First World urban experience, although of course, average conditions are clearly somewhat better in respect of the urban milieux. Further, in the Caribbean, this dovetailing of urban and rural is true whether we are talking of the behavioural realm of attitudes and ways of life, or structural conditions as observed on the ground. Certainly, in such contexts, Wirthian concepts of an entirely separate urban psyche appear to be somewhat dubious (Wirth, 1938). For example, the domination of land by the sugar plantations has meant that rural population densities are extremely high in Barbados. This is so due to the historical process described earlier, whereby in the immediate post emancipation era, as there was little land free for former slaves to cultivate, they had no choice but to labour on the plantations. In return, they were provided with rented house spots on marginal plantation or 'rab' lands. A series of acts kept the balance between landlord and tenant firmly in favour of the former, principally by keeping tenure highly insecure. Later, as Bridgetown grew and extended into what earlier had been plantation lands, much the same *tenantry* housing system was transferred to this urbanising scene. Thus, in both rural and urban areas insecurity of tenure related to the plantation system, and this has given rise to a remarkably uniform built landscape and vernacular architecture. The high density residential mosaic of Barbados is based on small wooden cabin houses, with the unit being set on a loose foundation of stones, so that should the eventuality arise, the house can be moved (see Potter, 1985a, 1986a). Thus, for the bottom two-thirds of the income spectrum of the population of Barbados, whether urban or rural, the form of vernacular architecture is strikingly uniform and it makes little sense to talk of urban and rural housing *per se*, although clearly the incidence of poor housing and poverty is greater in the rural areas of the country (see Potter, 1986b, p.196).

Urban-Rural Relations in the Contemporary Caribbean

Many of the features discussed above can be summarised to provide a highly simplified graphical representation of urban-rural interaction in small dependent islands of the Caribbean. This is shown in Figure 9.2. Obviously, the diagram depicts only some of the major internal flows which may be traced between what are conventionally referred to as 'rural' and 'urban' tracts of the national space, and the same selectivity has also been exercised with respect to the multiplicity of exogenous inflows and outflows which characterise developing countries.

The simple model shows the outflow of raw materials and agricultural staples to the metropoles, and the countervailing importation of foods, manufactures, new technology and the like. These are mediated via the linked processes of dependency and emulation, the consequences of which, following Frank and Beckford, need to be recognised not just in socio-economic terms, but also with respect to their deeply-ingrained psychological implications as well. But this framework has been sketched primarily with respect to the historical time period and the question remains, as to how satisfactory it is with regard to the types of socio-economic change which have occurred since the Second World War.

In this connection, there is a wide body of development theory, practice and observation serving to stress that the types of patterns established in the past have continued by and large to mould contemporary change and development. It can be argued that what Michael Lipton (1977, 1982) has described as a process of 'urban bias' came to characterise the search for post-independence development in the majority of Third World countries. Thus, it may be posited that Caribbean countries have exchanged colonialism for neo-colonialism in imitating the paths followed by First World countries during their industrial revolutions. The prescriptions of Rostow (1960) in his conceptualisation of *The Stages of Economic Growth: a non-communist manifesto* implied that the only barrier to 'development' was that of reaching a sufficiently high level of investment in one industrial sector of the economy to enable "take-off" to be reached. Thereby, the implicit argument appeared to run, 'backward' countries and regions would become urbanised,

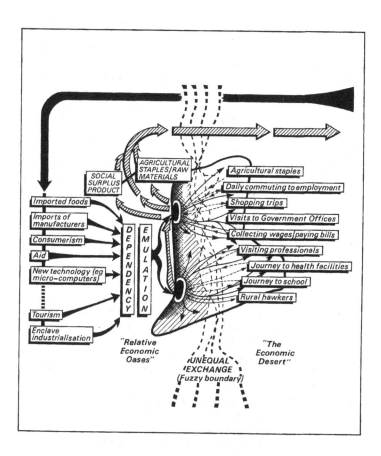

Figure 9.2: A simplified representation of urban-rural interactions in a small dependent economy

industrialised and thus "developed". In different fashions and with different emphases, the writings of Hirschman, Myrdal and Friedmann witnessed the first major struggles with these ideas. Would development spontaneously spill from the cores (urban centres) and into the peripheries (rural zones) (Friedmann, 1966), by virtue of 'trickling down' exceeding 'polarisation' (Hirschman, 1958); or would 'backwash' flows of factors of production and capital always tend to exceed the redistributive 'spread effects' and thereby stand perpetually in favour of urban-industrial zones (Myrdal, 1957)? Hoselitz (1955) had earlier aired the same problem primarily with respect to the precise role of the colonial city in the development process, coming to the conclusion that whilst cities induce parasitic interactions in the first instance, given time, their influence always changes to the benign generation of growth.

In policy terms, the post 1945 period witnessed these arguments becoming enshrined in a paradigm of development which is now referred to in shorthand terms as change "from above" or from the "centre downward" (see Friedmann and Weaver, 1979; Stöhr and Taylor, 1981; Potter, 1985a). Third World countries moved toward avowed policies of industrialisation (see Mountjoy, 1982b), first by import substitution and later by means of industrialisation through invitation. This approach was witnessed in the Caribbean by Puerto Rico's *Operation Bootstrap*, and was advocated in rounded policy terms by Sir Arthur Lewis (1950) for the Caribbean region as a whole. Barbados, and other Caribbean nations followed the advice, endeavouring to attract foreign manufacturers by tax incentives and the availability of low-cost labour. The programme of invited industry in Barbados, associated with the development of a series of manufacturing parks has been chronicled by Potter (1981). Whilst it is difficult to deny that such policies make sense for small dependent economies which desperately need to diversify their economies, the consequences are clear. New forms of dependency are the direct outcome, and multinationals become powerful agents in the shaping of the economy (Kowalewski, 1982). One of the reasons why such enterprises are attracted to developing countries is, of course, their relatively low wage levels. This factor, along with the relatively limited degree of training and skill enhancement received by operatives means that unequal exchange is perpetuated at the international level. Yet further, even in such small countries, the desire to impress and sway potential foreign investors is such that industrial parks come almost exclusively to be located in close proximity to the already existing facilities of the urban and suburban zones (see Potter and Hunte, 1979; Potter, 1981 specifically in relation to this point). Once again, the small size of Caribbean countries may well be envoked as a rationale for what is euphemistically described as a "central location", which geographically is anything but, and which thereby entails non-urban based workers undertaking formidable commuting trips on a daily basis if they are to be employed in the secondary sector of the economy.

Exactly the same centralised and unequal development process can be regarded as having characterised that other sector of post-1945 growth within the Caribbean region, and that is tourism. The geographical pull of safe and scenic beaches served to skew tourist development to those very same leeward coastal tracts that centuries earlier had first attracted mercantile capital. The concentration of present day social, recreational and transport facilities is a further factor tending to perpetuate the same pattern of polarised growth. On the other hand, rural poverty may in particular circumstances act as a clear negative externality as far as the tourist trade is concerned. Some of the complex social and economic nuances of the development of tourism have been outlined in respect of Barbados by Potter (1983a). Processes of unequal exchange are obviously involved in relatively lowly paid jobs such as waiting at table, domestic work, cooking, child minding, bar tending and the like. Yet other economic opportunities are afforded by informal sector and marginal activities including prostitution, especially of the male variety (see Karch and Dann, 1981). Again, geographical location and the size of the national territory frequently make it possible for hotel workers to commute to work from rural areas on a daily basis, and in so doing, lose out further in terms of the effective purchasing power of their hard-earned wages. Just as significantly, the economic multiplier associated with tourism is notoriously low, partly reflecting foreign interests in hotels and tourist facilities, but also the fact that tourism promotes massive imports, in particular of foods, furniture and consumer goods. Thus, tourism generally tends to further dampen and ultimately impoverish the indigenous agricultural sector. Emulation effects are likely to be such that a deepening of the damaging psychological effects of dependency occurs. The configuration of dependent urban-rural interactions emanating during the historical period has thereby tended to be intensified, and not transformed, by the twin contemporary trends of tourism and enclave manufacturing, and this is implied in Figure 9.2. Whilst it is not being argued that such developments are necessarily unacceptable, attention must be drawn to what would appear to be their inescapable consequences with regard to the urban-rural interface. Clearly, small dependent countries have little choice, and generally have to think in terms of taking what is on offer to them and endeavouring to adapt this to fit their needs in the most expedient manner. In the Caribbean, as elsewhere, the only other solution is to move toward greater regional economic integration. Here it is merely noted that these recent developments likewise serve to increase the differences between areas, and necessitate large volumes of spatial interaction between them in order to keep the unequal system 'ticking over'. If this type of interpretation is accepted, then some possible paths to change are implied, which whilst recognising the salience of dependent relations at all spatial levels, pushes toward national development via the consideration of priorities in an essentially integrated manner.

Rural-Urban Contrasts and Interactions: the case of Barbados

Urban-rural relations in the Caribbean are predicated on the dependency long engendered by the plantation system, and which has come to be reinforced in the era of political independence by the burgeoning of manufacturing, tourism and emulatory consumerism. It has also been suggested that small size means that it is possible to envisage rural-to-urban movements occurring on a daily basis, with rural folk being obliged to expend time, energy and money in travelling to the primate city or tourist zones to work, shop, hawk goods, pay bills, collect wages, visit doctors, places of entertainment and other required services (see Figure 9.2). Such moves are highly likely to occur mainly by public transport.

Having thus far made this point in general terms for Caribbean societies taken as a whole, in this final major section, the analysis presented above is given specific empirical expression via evidence relating to Barbados. This island, which is set apart, lying to the east of the main arc of southern Caribbean islands (Figure 9.3 inset), is home to some 248,983 people at a density of 578 per square km. The genesis of a highly uneven pattern of development in Barbados since its initial settlement by the British in 1627 has been outlined and analysed elsewhere by the present author (Potter, 1983b, 1985a, 1985b, 1986b, 1986c). The resultant highly focussed pattern of settlement and transport is summarised in Figure 9.3. Thus, a sharp divide separates the western and southern coastal strip which is made up of the urban parish of St. Michael, and the essentially high-status suburban parishes of Christ Church, St. James and St. Peter (Figure 9.3). These areas are by far and away the most affluent and contain much of the socio-economic infrastructure of the nation. On the other hand exist the remaining parishes of the east and north, and these are the main sugar producing rural-agricultural areas. Dann (1986) in a recent study has sub-divided these seemingly residual zones into what he refers to as 'rural 1' areas, those which are experiencing some growth, and 'rural 2' which are largely stagnant. The parishes of St. Philip and St. George are identified as belonging to the first group, whilst St. John, St. Joseph, St. Thomas, St. Lucy and St. Andrew make up the latter. This distinction is an interesting one, and notwithstanding the labels used, serves to emphasise the need to think in terms of quite subtle distinctions existing between what are all too frequently seen as entirely dichotomous boundaries (see Potter, 1986b specifically with relation to spatial inequalities in Barbados in this respect). In fact, in several of the data tabulations on urban-rural interactions presented below, these four area types will be used as categories.

The experiences of Barbados with respect to industrial development and tourism have already received some attention in this chapter. Here workplace movements and

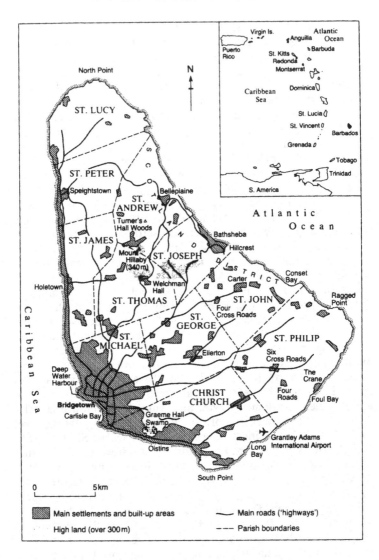

Figure 9.3: Barbados: principal settlements and roads (Source: Potter and Dann, 1987)

internal migration are first reviewed as they attest to the behavioural consequences of the type of dependency and polarised development discussed previously. Following this, the account turns to consider retail change and the shopping activities of a sample of households in rural, urban and suburban areas of Barbados. This section presents previously unpublished results obtained from a social survey carried out by the author during the summer of 1980. Finally, some data relating to the distribution of medical doctors and other essential services will be presented, which serve to show that urban-rural contrasts and resultant movements are endemic to all areas of life within this small dependent society.

Data collected as a part of the 1980/81 Population Census of the Commonwealth Caribbean provide some insights into the movement of people in connection with their employment. The situation is distilled in Table 9.1 which indicates the broad spatial location of the places of work of the economically active population of Barbados disaggregated by their parish of residence. The first observation is that leaving aside the metropolitan area of St. Michael, where 78.7 per cent of the resident workforce are employed in the home parish, generally, in excess of 40 per cent of the resident work force of each parish actually earns its living by travelling to and working in, the Bridgetown-St. Michael core area. This figure is particularly high for the predominantly rural inland parishes of St. George (49.2 per cent) and St. Thomas (48.2 per cent), the western most portions of which are increasingly becoming suburban commuter areas. But workplace dependency on St. Michael is also surprisingly high in the case of the rural 2 parishes of St. Thomas (48.2 per cent) and St. Andrew (33.3 per cent). This is also shown to be the case with regard to a number of relatively far distant rural areas, such as St. Joseph (45.3 per cent), St. John (40.3 per cent) and St. Philip (36.8 per cent). Of course, these high proportions reflect the concentrated spatial nature of employment in Barbados, where virtually all the non-agricultural jobs are located in St. Michael and the western coastal belt.

As would be anticipated, another area where significant rural-urban interaction occurs is with respect to residential migration. Some of the early work of the present author was specifically designed to determine how sharply divided were

Table 9.1: The working population of Barbados by parish of residence and location of work place

Area Type	Parish of residence	Percentage of resident population working in:		
		St Michael	Parish of residence	Other
Urban	St Michael	78.7	78.7	21.3
Suburban	Christ Church	47.0	42.3	10.7
	St James	44.1	39.2	16.7
	St Peter	26.8	44.8	28.4
Rural 1	St Philip	36.8	40.0	23.2
	St George	49.2	28.1	22.7
Rural 2	St John	40.3	36.7	23.0
	St Joseph	45.3	34.9	19.8
	St Thomas	48.2	30.8	21.0
	St Lucy	27.6	39.0	33.4
	St Andrew	33.3	44.4	22.3

the spatial perceptions of Barbadians concerning how desirable different areas of the national territory are from the point of view of wishing to live there (see for example, Potter, 1984). Preferences were shown to exist clearly in favour of the suburban coastal strip, that is specifically the parishes of Christ Church and St. James, and less so for the urban parish of St. Michael. Dann (1986) has recently calculated from the 1980/1 Census, statistics showing the actual migratory flows between the socio-geographical areas designated as urban, suburban, rural 1 and rural 2. The results are shown in Table 9.2. The most noticeable feature is the decline in the percentage of the resident population which stayed in the zone during the period 1970-80 as one moves from urban through to rural 2 areas. In fact, as high a proportion as 32.5 per cent and 26.5 per cent of the total populations of the rural 1 and rural 2 areas moved out respectively during the decade. In both cases, the preferred area for residence was, of course, the urban zone, accounting for a transfer of as many as 16.4 and 14.3 per cent of the total population of these zones respectively. The equivalent movements to suburban tracts was 9.2 per cent in the case of rural 2 areas and 7.7 per cent for the rural 1 parishes. There was also a movement of 12.4 per cent of the population of the suburban area during this time period, although this was almost entirely offset by a transfer of 11.9 per cent of the urban population to the suburban zone (Table 9.2). These statistics clearly reflect the existence of a strong divide between the designated socio-geographical area types and the continuation of a strong drift from the rural areas to the urban and suburban tracts.

We now turn to a further important area of spatial structure, and one which also serves to illustrate how recent developments have intensified dependent relations and unequal exchange, thereby increasing the potential for social welfare differences and rural-urban interactions. This relates to commercial change in Barbados, a topic which has recently been discussed by Potter and Dann (1986). The most salient fact concerning the present-day retail structure of the country is that a staggering 97 per cent of the total estimated national retail floorspace is to be found in the western and southern linear urban corridor, whilst this area accounts for only 77.93 per cent of national population and

Table 9.2: Internal migration in Barbados 1970–1980 by socio-geographical zone

| Area Type | Percentage of population remaining in zone | Percentage of Population migrating to: | | | |
		Urban zone	Suburban zone	Rural 1	Rural 2
Urban	83.9	–	11.9	2.5	1.7
Suburban	78.6	12.4	3.9	2.5	2.6
Rural 1	73.5	14.3	7.7	2.0	2.5
Rural 2	67.5	16.4	8.2	3.7	4.2

Source: Dann (1986) based on 1980/81 Census

54.02 of its land area. Some appreciation of the degree of concentration of retail infrastructure can be gained from Figure 9.4. Central Bridgetown alone accounts for 63 per cent of estimated national retail floorspace. The other major urban areas of Speightstown, Holetown and Oistins account for as little as 8.0, 4.6 and 2.7 per cent of national retail floorspace respectively.

The next most important retail areas tend to reaffirm the importance of the Bridgetown metropolitan area, for they comprise nine recently developed suburban commercial areas. In fact, these have all been built since 1965 and are based primarily on relatively large supermarkets. It is interesting to observe that the advent of one-stop supermarket shopping thereby coincided almost exactly with the expansion of mass tourism to the island in the early 1960s (Potter and Dann, 1986). Thus, the locations of these modern forms of retailing, with few exceptions, are related to outer metropolitan Bridgetown and the south-western coastal strip of tourist development. These supermarkets are effectively controlled by the 'Big Six' group of companies, which through intermarriage and corporatisation are responsible for virtually the entire commercial and mercantile activities of the island.

On the other hand, traditional forms of retailing in Barbados are represented by the *rum shop* and the *hawker*, or street vendor. Hawkers, generally poor rural Black folk have themselves been an important ingredient of rural-urban interaction and unequal exchange, carrying goods such as bananas, mangoes and the like to schools, places of entertainment, bus termini and the main streets of the city. In a similar fashion, rum shops remain as very important components of the retail system of Barbados, for many operators of such premises possess licences to also sell basic provisions (Potter and Dann, 1986). Although distributed throughout the country, such supply points are frequently the only ones existing in many rural villages of the east and north of the country (see Figure 9.4).

It may be suggested that retail change in Barbados has generally served to intensify the spatial polarity which characterises so many aspects of the socio-economic life of the country. Those people who live in the more disadvantaged parts of the island are faced with something

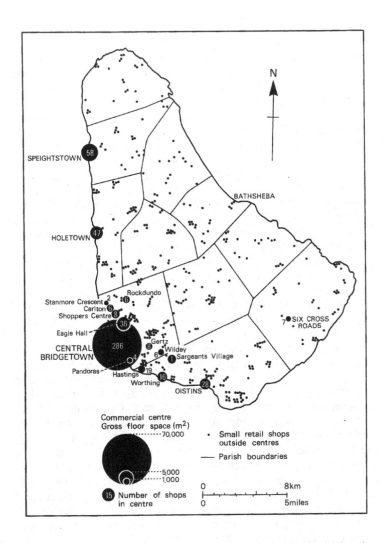

Figure 9.4: The spatial concentration of retail facilities in Barbados

of a dilemma. Either they can travel, generally by public transport, to avail themselves of the low prices and wide range of goods offered by the large metropolitan based supermarkets at the expense of time, effort and a two-way bus fare, or they can shop locally and face a very limited range of goods and inflated prices for those the prices of which are not controlled by the government.

Some indications of the resultant flow of rural denizens from the northern and eastern sections of Barbados is provided by the results of a social survey carried out by the author in 1980. In this work, a sample of 207 respondents drawn from the five settlements of Bridgetown, Speightstown, Oistins, Six Cross Roads and Belleplaine in approximately equal numbers were asked about their shopping- and employment-related travel activities. Most respondents reported that they visited Bridgetown quite frequently (Table 9.3). Residents of rural Six Cross Roads in the southeast reported that they travel there on average every 13 days. Even residents of the northeast at Belleplaine in St. Andrew visited the city centre every 28 days or so. Only denizens of the northern urban area of Speightstown were relatively immune to the attractions of the city centre, visiting it on average, approximately once in every 52 days. The principal reason for individuals travelling to the city centre from each of these rural or suburban environs was in each case to shop. The other major motive cited was in connection with business and employment, this accounting for just on 20 per cent of all respondents in the rural centres of Belleplaine and Six Cross Roads (Table 9.3). Other reasons explaining the movement of rural residents to the city are just as interesting, although accounting for only relatively small numbers. Thus, 13.3 per cent of the Six Cross Roads sample visited Bridgetown in order to pay bills and shop, whilst 4.4 per cent reported that they travelled to the city in order to 'lime', that is to hang around and meet friends. For the respondents based in Belleplaine, some 7.32 per cent travelled in principally to pay their bills, and as many as 7.31 per cent did so in order to collect their pay (Table 9.3).

The interviewees were also asked where they normally shopped for three broad categories of merchandise: bread, main provisions and clothing. The shopping patterns thus

Table 9.3: Frequency of visits to Bridgetown among residents of five areas of Barbados and the principal reason for making such trips

Survey Location	Mean number of days between visits to Bridgetown	Percentage of respondents stating principal reason for trip as being:							
		Shop	Business Employ't	Shop/ Entertain-ment	Shop/ Other	Pay Bills /Shop	Pay Bills	Collect pay	'Lime'	Other/ NR
Bridgetown	11.20	55.56	24.44	0.00	6.67	0.00	4.44	0.00	0.00	8.89
Speightstown	51.68	51.61	9.67	22.58	0.00	3.23	0.00	0.00	0.00	12.91
Oistins	18.67	60.00	13.33	6.67	0.00	4.44	4.44	0.00	2.22	8.90
Six Cross Roads	13.44	33.33	20.00	0.00	17.79	13.33	0.00	0.00	4.44	11.11
Belleplaine	28.03	41.46	21.95	7.32	9.76	2.44	4.88	7.31	0.00	4.88

Source: Author's survey (1980)

revealed among the respondents are summarised in Table 9.4. The most noticeable feature is the heavy dependence of the Belleplaine sample on Bridgetown for purchases of even low-order goods such as bread and main provisions, with the city accounting for 39.02 and 46.34 per cent of purchases in these two categories respectively. Although much lower, 17.78 and 11.11 per cent of the Six Cross Roads sample shopped for bread and provisions in Bridgetown respectively. While, in part, these figures reflect the levels of journeying into Bridgetown for reasons such as employment, paying bills and the like as discussed above, they attest once more to the very high overall rates of rural movement into the city. Again residents of Speightstown are the only ones who are relatively independent of the city centre, apart that is, for purchases of clothing (Table 9.4). Here, just over half of those interviewed from Speightstown reported that they travelled to Bridgetown. But this proportion is far higher for the other geographically defined samples, exceeding 90 per cent in the case of Six Cross Roads and 78 per cent for Belleplaine.

This sample analysis provides some quantitative expression of the heavy dependence of the rural residents of Barbados on Bridgetown for even quite humble, every-day purchases. The rural-urban flows which are thus generated reflect uneven social provision in a small dependent country, and they affect most other areas of daily life as well. For instance, Dann (1986) notes how a recent national telephone directory reveals that 80 per cent of Government ministries, 60.5 per cent of branch banks, 94.5 per cent of insurance companies, and 100 per cent of lawyers are to be found located in the wider Bridgetown urban zone. Thus, when rural residents need to make use of such services, "then they must subject themselves to the rigours and unreliability of public transport" (Dann, 1986, p.13). The inverse care law would appear to be no less true of the provision of medical facilities, at least as expressed in the geographical distribution of medical practitioners. The present author found 166 medical doctors listed in the 1983-84 Barbados Telephone Directory. These were allocated to the four socio-geographical zones used previously in this section, and the results are shown alongside percentage population shares for the same areas in Table 9.5. The urban area of St.

Table 9.4: Shopping patterns among residents of five areas of Barbados

(a) Shopping for Bread:

Survey Location	Percentage of respondents shopping in:		
	Bridgetown	Home Parish	Other/NR
Bridgetown	88.89	88.89	11.11
Speightstown	6.45	67.74	25.81
Oistins	11.11	68.89	20.00
Six Cross Roads	17.78	60.00	22.22
Belleplaine	39.02	43.91	17.07

(b) Shopping for Main Provisions:

Survey Location	Percentage of respondents shopping in:		
	Bridgetown	Home Parish	Other/NR
Bridgetown	88.89	88.89	11.11
Speightstown	9.68	67.74	22.58
Oistins	22.22	53.34	24.44
Six Cross Roads	11.11	60.00	28.89
Belleplaine	46.34	26.83	26.83

(c) Shopping for Clothes:

Survey Location	Percentage of respondents shopping in:			
	Bridgetown	Home parish	Overseas	Other/NR
Bridgetown	75.56	75.56	11.11	13.33
Speightstown	54.84	32.26	12.90	0.00
Oistins	66.67	2.21	15.56	15.56
Six Cross Roads	91.12	2.22	4.44	2.22
Belleplaine	78.06	2.43	9.76	9.75

Source: Author's survey (1980)

Table 9.5: Medical doctors listed in the 1983-84 Barbados Telephone Directory according to Parish

Socio-Geographical zone	Number of doctors listed	Percentage of total doctors	Percentage of total population
Urban	95	57.3	40.2
Suburban	52	31.3	27.6
Rural 1	10	6.0	14.5
Rural 2	9	5.4	17.7

Michael accounts for well over half the doctors to be found within the country, although containing only 40 per cent of its population. The suburban areas show a broad equivalence between doctors and population, although being tipped slightly on the favourable side. But the situation with regard to the rural tracts of Barbados is far different. Thus, both rural 1 and rural 2 areas, but particularly the latter, are shown to be disadvantaged with regard to health care facilities, suggesting that they must vie with a greater number of fellow patients, or alternatively again have to consider the option of travelling in order to visit doctors who are situated outside their immediate geographical localities.

Concluding Comments

In concluding, we return to some of the themes introduced at the beginning of this chapter. The first point is that 'urban' and 'rural' areas, in so far as they can be recognised at their extremes as different in the Caribbean, are so differentiated not because they are affected by separate sets of processes, nor by the same processes acting to contrasting degrees, but rather, because they have been influenced by a single process of dependent development. It is hoped that this chapter has gone some way toward showing that the relations existing between such areas are so complex that even greater caution must be exercised when using the labels 'urban' and 'rural' in the Caribbean than is normally the case elsewhere; unless that is, it is clearly acknowledged that the terms merely represent convenient shorthand descriptors of areas which a single process of uneven socio-spatial development has served to differentiate markedly.

These arguments also suggest the pressing need for the adoption of an holistic approach in studies of Caribbean development and social change. It may be ventured that the geographical perspective is particularly well-suited and necessary in this connection. It is somewhat of a paradox that in small countries, where an holistic view is so important, both with regards to analysis and policy formulation, that small size may mean that spatial and geographical facts seem to be taken-for-granted (Potter and Dann, 1986). It is this reality which sometimes leads to an overemphasis being placed on national economic planning. Due recognition, however, must be accorded integrated policies which include the needs of all areas of the national territory. Whilst internationally the clarion call may well be for 'development from below', employing agropolitan principles (Stöhr and Taylor, 1981; Friedmann and Weaver, 1979), it must be appreciated that small dependent Caribbean countries are not able to cut themselves off in the way that China has done from time to time. But development in the region must increasingly stress *basic needs* and the social welfare of rural-peripheral residents, in order to raise their standards of living, but more saliently, their overall quality of life. Accordingly, there is much in the plea for the pursual of integrated rural-urban programmes of development in the Caribbean (see Gomes, 1985). The issue is best seen as one of reducing exploitative interactions based on unequal exchange. Thus, for example, in the Caribbean context, whilst enclave manufacturing has to be accepted as an economic necessity, rural locations should be stressed. Likewise, tourism might be used to encourage the development of small-scale indigenous family businesses, an aim frequently espoused, but seldom if ever actually implemented. All in all, therefore, *social criteria* must be stressed as much, if not more than economic ones. A principal aim must be to control the rate at which Third World patterns of urban consumption converge on those of First World nations, for this can only lead to further socio-spatial inequalities if it occurs in advance of genuine rural development (Armstrong and McGee, 1985). Major changes in perceptions are required if dependency is to be reduced, as noted by Beckford (1972) in his emotive call to decolonise the mind. Perhaps at first these changes can best be stimulated by giving specific prominence to the importance of indigenous rural-based strategies of development and change.

Acknowledgements

At the time of writing, the author was in receipt of a grant from the Nuffield Foundation, in support of a project examining housing, planning and development in the Caribbean. Tony Binns, Lennox Britton, Graham Dann and

Tim Unwin were kind enough to read and comment on the chapter. Graham and Elizabeth Dann's hospitality and support during fieldwork is recorded with gratitude.

Rural-Urban Interaction in the Caribbean

REFERENCES

Armstrong, W. and T.G. McGee (1985) *Theatres of Accumulation: Studies in Asian and Latin American Urbanization*, Methuen, London

Beckford, G.L. (1972) *Persistent Poverty: Underdevelopment in Plantation Economies of the Third World*, Oxford University Press, New York

Beckford, G.L. (1975) 'Caribbean rural economy', in: Beckford, G.L. (ed.), *Caribbean Economy: Dependence and Backwardness*, Institute of Social and Economic Research, Mona, Jamaica

Blume, H. (1974) *The Caribbean Islands*, Longman, London

von Boventer, E. (1962) 'Towards a unified theory of spatial economic structure', *Papers of the Regional Science Association* 10, 163-87

Brookfield, H.C. (1975) *Interdependent Development*, Methuen, London

Clapp, J.M. and H.W. Richardson (1984) 'Technological change in information processing industries and regional income differentials in developing countries', *International Regional Science Review* 9, 241-56

Clarke, C.G. (1974) 'Urbanization in the Caribbean', *Geography* 59, 223-32

Clarke, C.G. (1975) 'The Commonwealth Caribbean', in: Jones, R. (ed.), *Essays on World Urbanization*, George Philip & Son Limited, London

Clarke, C. (1986) 'Sovereignty, dependency and social change in the Caribbean', in: *South America, Central America and the Caribbean*, Europa Publications, London

Cross, M. (1979) *Urbanization and Urban Growth in the Caribbean: an Essay on Social Change in Dependent Societies*, Cambridge University Press, Cambridge

Dann, G.M.S. (1986) 'Some observations on the nature of Barbadian society, and its capacity to cater to the needs of the mentally ill and elderly', *unpublished paper prepared for the Pan American Health Organisation*, 35pp.

El-Shakhs, S. (1972) 'Development, primacy and systems of cities' *Journal of Developing Areas* 7, 11-36

Frank, A.G. (1969) *Capitalism and Underdevelopment in Latin America*, Monthly Review Press, New York

Friedmann, J. (1966) *Regional Development Policy: a case study of Venezuela*, M.I.T. Press, Cambridge, Massachusetts

Friedmann, J. and C. Weaver (1979) *Territory and Function: the Evolution of Regional Planning*, Arnold, London

Gilbert, A. (1985) *An Unequal World: the links between Rich and Poor Nations*, Macmillan Education, Basingstoke and London

Girvan, N. (1973) 'The development of dependency economies in the Caribbean and Latin America: review and comparison', *Social and Economic Studies* 22, 1-33

Gomes, P.I. (ed.) (1985) *Rural Development in the Caribbean*, Hurst, London and St Martin's Press, New York

Graham, G.E. and I. Floering (1985) *The Modern Plantation in the Third World*, Croom Helm, London

Harvey, D. (1973) *Social Justice and the City*, Arnold, London

Harvey, D. (1985) *The Urbanization of Capital*, Blackwell, Oxford

Hirshman, A.O. (1958) *The Strategy of Economic Development*, Yale University Press, New Haven

Hope, K.R. (1983) 'Urban population growth in the Caribbean', *Cities* 1, 167-74

Hope, K.R. (1986) *Urbanization in the Commonwealth Caribbean*, Westview Press, Boulder

Hoselitz, B.F. (1955) 'Generative and parasitic cities', *Economic Development and Cultural Change* 3, 278-94

Karch, C. and G. Dann (1981) 'Close encounters of the Third World', *Human Relations* 34, 249-68

Kowalewski, D. (1982) *Transnational Corporations and Caribbean Inequalities*, Praeger, New York

Lewis, W.A. (1950) 'The industrialisation of the British West Indies', *Caribbean Economic Review* 2, 1-61

Lipton, M. (1977) *Why Poor Poeple Stay Poor: Urban Bias in World Development*, Temple Smith, London

Lipton, M. (1982) 'Why poor people stay poor', in: Harriss, J. (ed.), *Rural Development: theories of peasant economy and agrarian change*, Hutchinson, London

Lösch, A. (1940) *The Economics of Location*, Yale University Press

Lowenthal, D. (1972) *West Indian Societies*, Oxford University Press, London

Mountjoy, A.B. (1976) 'Urbanization, the squatter and development in the Third World', *Tijdschrift voor Economische en Sociale Geografie* 67, 130-37

Mountjoy, A.B. (1978) 'Urbanization in the Third World' in: Mountjoy, A.B. (ed.), *The Third World: problems and perspectives*, Macmillan, London

Mountjoy, A.B. (1982a) 'Squatters in Hong Kong', *Geographical Magazine* 53, 119-25

Mountjoy, A.B. (1982b) *Industrialisation and Developing Countries* (5th edition), Hutchinson, London

Myrdal, G. (1957) *Economic Theory and Underdeveloped Areas*, Duckworth, London.

O'Connor, A. (1983) *The African City*, Hutchinson, London

Palmer, R.W. (1979) *Caribbean Dependence on the United States Economy*, Praeger, New York

Potter, R.B. (1981) 'Industrial development and urban planning in Barbados', *Geography* 66, 225-28

Potter, R.B. (1983a) 'Tourism and development: the case of Barbados, West Indies', *Geography* 68, 44-50

Potter, R.B. (1983b) 'Congruence between space preferences and socio-economic structure in Barbados, West Indies', *Geoforum* 14, 249-65

Potter, R.B. (1984) 'Mental maps and spatial variations in residential desirability: a Barbados case study', *Caribbean Geography* 1, 186-97

Potter, R.B. (1985a) *Urbanization and Planning in the Third World: Spatial Perceptions and Public Participation*, Croom Helm, London and Sydney and St. Martin's Press, New York

Potter, R.B. (1985b) 'Environmental planning and popular participation in Barbados and the Eastern Caribbean: some observations', *Bulletin of Eastern Caribbean Affairs* 11, 24-30

Potter, R.B. (1986a) 'Housing upgrading in Barbados, West Indies: the Tenantries Programme', *Geography* 71, 255-57

Potter, R.B. (1986b) 'Spatial inequalities in Barbados, West Indies', *Transactions of the Institute of British Geographers*, New Series 11, 183-98

Potter, R.B. (1986c) 'Physical development or spatial land use planning in Barbados: restrospect and prospect', *Bulletin of Eastern Caribbean Affairs* 12, 24-32

Potter, R.B. and Binns, J.A. (1988) 'Power, politics and society' in: Pacione, M. (ed.), *Geography of the Third World - Progress and Prospects*, Routledge, London and New York

Potter, R.B. and Dann, G.M.S (1986) 'Core-periphery relations and retail change in a developing country: the case of Barbados, West Indies', *Paper presented at the Symposium on Commercial Change, International Geographical Union Study Group on the Geography of Commercial Activities*, Barcelona, Spain

Potter, R.B. and G.M.S. Dann (1987) *Barbados (World Bibliographical Series Vol.6)*, Clio Press, Oxford

Potter, R.B. and M.L. Hunte (1979) 'Recent developments in planning the settlement hierarchy of Barbados: implications concerning the debate on urban primacy', *Geoforum* 10, 355-62

Roberts, B.R. (1978) *Cities of Peasants: the Political Economy of urbanization in the Third World*, Arnold, London

Rostow, W.W. (1960) *The Stages of Economic Growth: a non-communist manifesto*, Cambridge University Press, London

Stöhr, W.B. and D.R.F. Taylor (1981) *Development from Above or Below? The Dialectics of Regional Planning in Developing Countries*, Wiley, Chichester.

United Nations (1975) *Trends and Prospects in Urban and Regional Population, 1950-2000*, United Nations, New York

Vance, J.E. (1970) *The Merchant's World: the geography of wholesaling*, Prentice-Hall, Englewood Cliffs

Webber, M.J. (1971) *The Impact of Uncertainty on Location,* M.I.T. Press, Cambridge, Massachusetts

Wirth, L. (1938) 'Urbanism as a way of life', *American Journal of Sociology* 44, 1-24

10

A PERIPHERY IN GENESIS AND EXODUS: REFLECTIONS ON RURAL-URBAN RELATIONS IN JAMAICA

DAVID BARKER

Rural-urban interaction in developing countries takes a variety of forms and occurs in the context of stark differentials in lifestyles, income-generating opportunities and service provision between rural areas and urban centres. Concepts such as pole and periphery can help us analyse the structure of rural-urban relations, and to understand how regional differentials can get worse rather than better. The most striking examples of rural-urban imbalances can be found in Latin America, Africa and Asia, in countries which are large political entities, both in geographical area and population. There have been fewer studies of rural-urban relations in regions composed of small island states.

Caribbean islands, especially the larger ones in the Greater Antilles, have many features typical of rural-urban imbalance; rural peripheries characterised by low incomes and out-migration, urban primacy and the proliferation of informal sector activities and the spatial and socio-economic concentration of wealth and decision-making. Yet, today, the majority of rural areas in the Caribbean are, at most, only a few hours travel time from their capital by public transport, although Haiti is an exception. The small areal size of an island, a small population and a limited endowment of resources significantly temper the form, pattern and process of rural-urban relations.

This essay traces the origins and emerging patterns of rural-urban contact in Jamaica. Though much bigger than many eastern Caribbean territories, the country is small by Third World standards, with a land area of only 4,411 square miles. However, its capital, Kingston, has a population approaching 700,000 in a total national population of 2.3 million, and so it is appropriate to use the concept of pole-periphery in a discussion of the unfolding spatial organisation of rural and urban areas.

In briefly reviewing aspects of rural-urban interaction in Jamaica, three factors emerge which are relevant to a more general view of rural-urban relations in small tropical island regions like the Caribbean. The relative importance of these factors is, perhaps, somewhat different compared to those which might be identified in equivalent studies of rural-urban interaction in larger developing countries.

The first of these factors is that the limited areal size of an island heightens competition for scarce land resources. The historical and modern development of the rural space economy in Jamaica has been dominated by competition for the limited supply of agricultural land, between competing social groups of unequal size and power. A dualist structure emerged in the colonial plantation economy, characterised by a traditional, small-scale farming sector geared largely to domestic food production, and a more modern, large-scale sector, mainly export-oriented. This structure had economic, socio-cultural, political and spatial connotations. The resultant land use patterns have given a distinctive signature to Jamaica's landscape. Full slave emancipation in 1838 was instrumental in creating this dualism, and it also triggered a widespread redistribution of the country's rural population. The date serves as a benchmark for comprehending the high levels of geographical mobility of the Jamaican population, both historically and in modern times.

A second factor which heavily circumscribes rural-urban relations in the Caribbean is the dominating influence of migration. It is the persistence and magnitude of migration in relation to total national population sizes that is significant. Information flows created in the aftermath of migration underpin rural-urban contact and perceptions. In Jamaica, internal migration streams appeared after emancipation, but were given form, direction and substance by the twentieth century emergence of Kingston as a primate city, and later, by the rapid post-1950 development of bauxite and tourism. International migration also had nineteenth century antecedents, as emancipation opened up the prospect of movement overseas, but in the twentieth century, it has become endemic. Current estimates suggest that more than 2.6 million people who were born in Jamaica now live abroad; a figure which is higher than the total national population. There are large Jamaican communities, and their children and grandchildren, living in the towns and cities of Britain, Canada and the United States, with over half a million Jamaicans in New York City alone. The Jamaican extended family, like its counterpart in other Caribbean countries, has become international. There is considerable, regular two-way traffic between family members at home in Jamaica and abroad, adding a cosmopolitan dimension to the information flows which affect the arena of rural-urban contact.

A third factor of considerable importance in examining rural-urban interaction in the Caribbean is the environment. Physically, small tropical islands consist of fragile interlocking sets of delicately balanced ecosystems. An island's interior is in close spatial proximity and has intimate environmental linkages with, the coastal zone. This is especially important in islands with mountainous interiors, such as those in the Greater Antilles and Windward Islands. Pressure on land resources in the rural interior can have ecological as well as economic repercussions on urban settlements, many of which are located on or near the coast. A study of rural-urban interaction on small tropical islands, therefore, needs to take into account the environmental reciprocity between the rural interior and coastal urban centres.

The first part of this chapter examines the origins of the rural space economy in Jamaica, as a key to understanding the structure of the modern rural periphery. The second section describes the development of the metropolitan core, focussing on the rise of Kingston as a primate city. It also documents the emergence of growth poles associated with the development of bauxite and tourism. A final section focusses on the rural periphery, but its main purpose is to illustrate how environmental factors need to be accommodated in order to obtain an holistic view of relations between the urban core and rural periphery.

The Evolution of the Rural Space Economy in Colonial Jamaica

Jamaica was a Spanish colony for over 150 years after it was first visited by Columbus during his second voyage in 1494. The Spanish enslaved and eventually exterminated the indigenous Arawak population, and imported small numbers of African slaves. However, Spain did not devote much time to her colony, and the activities of the small number of Spanish settlers were mainly confined to cattle rearing on the southern plains, around their capital Villa de la Vega (now called Spanish Town).

The English occupied the island from 1655. The Spanish fled to Cuba, but some of their former slaves founded renegade maroon communities in the isolated mountains at the eastern and western ends of the island. Their numbers were continuously replenished by escapees from English plantations, and the maroons distinguished themselves in minor "wars" and frequent skirmishes against the British (Eyre, 1980). Details of agricultural practices in their villages are sketchy, but they seem to have practised a type of bush fallowing, and some aspects of their livelihood reflected strong African roots (Kopytoff, 1973; Barker and Spence, 1988). Though small in numbers, their modern descendants enjoy symbolic respect as long-established, independent farming communities.

Jamaica began to assume the classic characteristics of a plantation economy once it became an English colony. Early plantation crops were cocoa and pimento, but sugar rapidly asserted itself as the island's principal crop. Higman (1987) records the rise in the number of sugar plantations from 57 in 1670, to 419 in 1739 and 1,061 in 1786. In 1730, Jamaica was the major sugar producer in the British empire, and for a period was Britain's richest colony. Sugar production peaked in 1805 at almost 100,000 tons, a figure which was not achieved again until 1936.

The key factors in sugar production were extensive flat land and good soils, and slave labour. The British plantation owners, therefore, chose the best agricultural lands for their estates, and these were located on the coastal plains around the island's perimeter and on the wide fertile inland valleys, or poljes, located in limestone regions. Worthy Park estate in

St. Catherine is an example of the latter which has enjoyed over 300 years of continuous sugar production. Although sugar plantations dominated the economy, there were other types of plantations and estates. Higman (1976) notes that in 1832, only half Jamaica's slave population were to be found on sugar plantations. Among a variety of other enterprises, coffee plantations accounted for 14 per cent of the slave population and livestock pens for another 13 per cent. Coffee plantations were located in upland regions, notably the Blue Mountains in the eastern part of the island and in the parish of Manchester (Higman, 1986). Blue Mountain coffee was of high quality and much sought after in the English coffee houses of the eighteenth century. It still enjoys a reputation as one of the World's finest and most expensive coffees. Cattle pens were located on the coastal lowlands and also in the flatter uplands of St. Ann parish.

Plantation owners found it difficult and expensive to import food for their labour forces because of the frequent wars which punctuated the Caribbean. They therefore allowed their slaves to grow their own food, but imposed severe constraints on this, in terms of the amount of time allowed for cultivation, and the type of land the slaves could use.

Thus slaves were only allowed to cultivate their own plots on Saturday afternoons, Sundays, and at holidays. They cultivated two types of plots. One was a tiny area around their cabins on the estates, devoted mainly to tree crops such as avocado, mango, coconut, banana, and plaintain, with a few root crops, herbs and spices. These were undoubtedly the forerunners of the ubiquitous Caribbean kitchen gardens (Brierley, 1976), or to use an alternative name, food forests (Hills and Iton, 1983). The other type of plots were called "provision grounds" or "polinks" (Mintz, 1974). These plots were generally on unused plantation hillsides and had poor soils, since the best land was monopolised by the principal economic activities of the plantations. Provision grounds were also marginal in location; slave villages tended to be centrally located on plantations and therefore provision grounds might be several miles away. The main produce cultivated by slaves on this type of plot were root crops such as yams, sweet potatoes and dasheen, which are still known today as "ground provisions".

The production of food by plantation slaves seems to have been highly successful, despite the geographical stranglehold which the plantations had on the best land, the hardships imposed by the terrain which had to be cultivated, and the limited time available to cultivate their own fields after gruelling labours of the plantation were finished for the week. Mintz (1985, p.134) comments: "were it not such a deplorable comment on slavery itself, it would be funny to realize that the British garrisons maintained in Jamaica before emancipation 'to keep order' were almost entirely provisioned by the slaves, working on their own time. One writer, an enthusiastic supporter of slavery at that, concluded that perhaps as much as twenty per cent of the metallic currency in Jamaica at the end of the eighteenth century was in the hands of the slaves." He went on to observe that the island's free population were also dependent on the food which had been grown by slaves working for themselves. Fruit and vegetables were available largely through a thriving system of "native markets" which had been allowed to develop across the island by the latter part of the eighteenth century.

Held on Sundays, these markets became important arenas for economic and social contact between rural and urban-based slaves. A marked sexual division of labour in marketing agricultural produce was apparent even at this time, since vending was dominated by women. The tradition is directly related to West African ancestry, and the higglers themselves would have been a mixture of creole and African-born slaves. Simmonds (1987) notes that urban slave higglers were allowed considerable geographical mobility to engage in itinerant vending in rural areas. This, too, would have created new avenues for the exchange of goods and produce, and news and information, between town and country. The foundations of Jamaica's famous higglering system, an important facet of modern rural-urban contact, were laid in the pre-emancipation period.

Full emancipation, in 1838, is widely acknowledged as the genesis of the peasantry in the Anglophone Caribbean (Marshall, 1972). Mintz (1985) refers to earlier groups of small farmers, such as the maroons, as a proto-peasantry. Emancipation permitted the first free geographical movement for the island's 311,000 slaves out of the national population of 370,000. The majority of these slaves lived in compact villages on plantations, and not surprisingly, they left the plantations in their thousands. Thus, emancipation triggered a significant redistribution of the island's rural population. The majority of ex-slaves settled Jamaica's upland areas and began to establish an independent agrarian life and a new rural order. Some left the plantations voluntarily, others were forced to move because their former masters imposed high rentals for the use of housing and provision grounds on plantation land. The period of movement and readjustment was gradual, over a decade or so, and some ex-slaves chose to sell their labour to the plantations, working either on a full-time or seasonal basis.

The move away from the plantations created serious labour shortages for their owners. Large numbers of indentured labourers were imported from various countries over the next eighty years to try and solve the problem. The largest group came from the Indian sub-continent, but others were brought from China, Africa and Europe. Despite these efforts, plantation agriculture went into general decline, sugar production was particularly affected by the Sugar Equalisation Act of 1846, which ended preferential treatment for sugar imported into Britain from her colonies. Many plantations were subdivided and sold off as freehold land or abandoned. Higman (1986) reported that between 1832 and 1847, 465 coffee plantations were abandoned, representing 488,000 acres. The number of sugar estates declined from 670 in 1834 to 200 in 1880 (Higman, 1987). Other plantations sold off their marginal lands (often provision grounds) in order to stay viable.

Basically, the redistribution of population created a dispersed rural settlement pattern across Jamaica, and opened up new areas to agriculture. In addition, there were many new clustered communities formed which were founded as "free villages". They were established by religious missionary groups such as the Baptists. The first free village was Sligoville, in the hills above Spanish Town, but the majority were located in the northern parishes. Free Hill, for example, in the parish of St. Ann is symbolically perched in the hills overlooking the former sugar estates of Richmond and Llandovery, which are over a thousand feet below on the narrow coastal plain.

Rural-Urban Relations in Jamaica

Nineteenth century Jamaica was an agrarian society and at the time of emancipation its towns were mainly small coastal ports (Figure 10.1), which functioned as gateways for imports and exports. Kingston was easily the largest town although it did not become the capital until 1872. Some former slaves migrated to these towns after emancipation. Although the numbers were probably fairly small, the island-wide movement from rural areas to towns has continued ever since. As early as the 1850s, Jamaicans were also migrating overseas. An estimated 5,000 went to Panama to construct the Panama railway. Senior (n.d.) reports they came from all walks of life, but most were living in Kingston and Spanish Town at the time of their move. Fifty years later, another wave of emigrants went to the same place to work on the construction of the Panama Canal.

Another effect of the redistribution of the rural population was to give an impetus to the growth of inland market centres. Small villages such as Linstead and Brown's Town began to grow into market towns and assert their functional importance in their relations with the surrounding countryside. The coastal towns continued to have their own important produce markets and it was in the post-emancipation period that the country higgler came to epitomise the economic, social and cultural significance of Jamaica's markets as points on contact between town and country people.

Thus, the net result of the redistribution of the rural population in the post-emancipation period was to create a broad structural dichotomy in the rural space economy. It was fractioned into a pattern of large-scale estates and small-scale farms, which occupied separate geographical locales. The plantations, estates and pens were located on the flatter, alluvial coastal plains and inland valleys, with some exceptions, such as coffee and pimento growing districts and the cattle pens of St. Ann. Small farms had a more general island-wide distribution (in inhabited areas), but the great majority were located in the upland interior.

Although plantation agriculture went into a period of decline in the second half of the nineteenth century, it retained the best agricultural land. Planters who remained in production held on tenaciously to their land, selling off poorer marginal land only when absolutely necessary. The

301

Figure 10.1: Principal settlements and bauxite areas of Jamaica

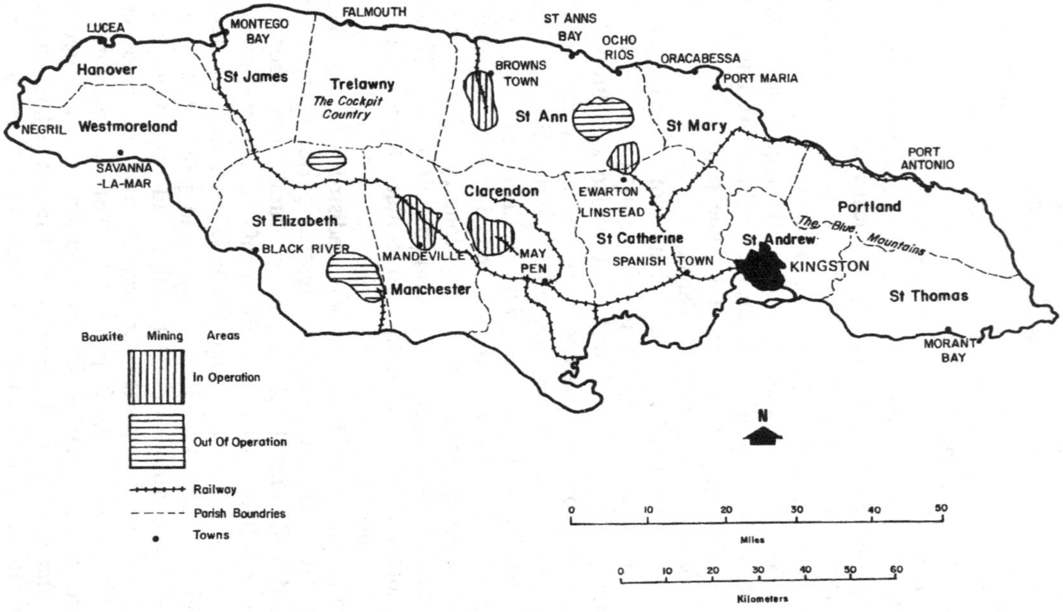

302

peasantry bought, leased and rented land, but the land hunger was so great that land which appeared to be unoccupied was often 'captured' (i.e.squatted). Insecurity of tenure and land hunger were important grievances that fuelled the Morant Bay rebellion of 1865. The rebellion was savagely suppressed, but afterwards the Planter dominated Assembly was dissolved and Crown Colony rule from Westminster established.

The majority of small farmers engaged in subsistence farming, using bush fallowing methods. Land was cleared with the help of fire, a technique which was known as 'fire stick' cultivation at the time. The pattern of land holdings was small and fragmented, and farming steep hillsides was arduous and relatively unrewarding. In time a unique land tenure system, called "family land" became an integral part of the system in Jamaica (Besson, 1984) and also in the rest of the Caribbean.

Despite the subsistence orientation to small farming, it was noted above that there was a sufficiently large surplus to support a thriving internal marketing system. Small- and medium-sized farms also began to grow export crops such as cane and coffee. More significant, however, was the rapid growth of the banana industry, which attracted many small-scale farmers. Bananas had been cultivated in Jamaica since the earliest European colonisation, but it was a variety introduced from Martinique in 1835, called Gros Michel, which attracted the export trade.

The first export cargo of Jamaican bananas was shipped from Orcabessa and Port Antonio to Boston by Captain George Busch in 1866. The banana trade began in earnest in 1870 when another enterprising sea captain, Lorenzo Dow Baker, shipped bananas from Port Morant. Port Antonio, Monego Bay, Lucea, Falmouth, St Anne's Bay and Port Maria all became important banana ports on the north coast. Indeed, Port Antonio's population grew from 1,700 in 1891 to 7,000 in 1911, rivalling the population of Spanish Town, Jamaica's second largest town.

Bananas became Jamaica's "green gold" and the Gros Michel variety acquired the dialect name of "goyark" in the 1890s, signifying an intent to ship the fruit to New York. In the twentieth century, Jamaica became Britain's principal source of bananas. Yet rural prosperity remained elusive for the smaller banana growers. The industry was dominated by powerful companies such as the United Fruit company who, in addition to their shipping interests, quickly acquired large land holdings and established commercial banana estates. The Jamaica Banana Growers Association was not formed until 1929 to protect the interests of local farmers.

The entry of small farmers into the production of cash export crops like bananas, cane and coffee further blurred the strict dichotomy between a large-scale export oriented sector and a small-scale domestic market sector. However, the locational patterns which characterised the two components remained largely unaltered.

Jamaica's population grew from 441,000 in 1861 to 639,000 in 1891, and to 858,000 by 1921. The major part of this increase took place in the rural areas. Kingston's population was 34,000 in 1861, representing approximately 13 per cent of the national population. Between 1891 and 1921, Kingston and its adjoining urban areas on the Liguanea Plain had increased its population from 58,000 to 89,000, yet this constituted a decline in terms of the proportion of national population to 11 per cent, and then under 10 per cent, respectively. Lewis (1973) notes that between 1882 and 1896 the number of small farmers with less than 10 acres increased from 52,000 to 81,900. Thus, as the rural agricultural population began to grow, the impact of population pressure in hillside farming areas began to be felt. Eyre (1974) traces the spatial pattern of rural population dynamics after 1911, and documents the increasing number of rural districts with population densities greater than 250 per square mile. A concomitant of population pressure on land resources was a growing exodus of people away from the rural areas: rural-urban migration began to gather pace in the twentieth century.

The emergence of an urban core and regional growth poles

Jamaica's economy began to assume a more modern urban-industrial character as the twentieth century progressed. As the spatial organisation of the economy evolved, there was an increasing polarisation between a rapidly expanding urban core and a lagging rural periphery. A major motive force in this spatial reorganisation was out-migration, from rural areas to urban destinations in Jamaica and abroad. In this section the development of the urban-industrial economy is examined, in the context of rural-urban interaction. The most significant phenomenon in the first half of the twentieth century was the emergence of Kingston as a dominating primate city. More recently, patterns of rural-urban interaction have become complex with the appearance of small growth poles associated with the bauxite and tourist industries.

Kingston was founded on the Liguanea Plain after an earthquake had destroyed the rich and infamous buccaneer town of Port Royal in 1692. It was noted earlier that Kingston was Jamaica's largest town even before it became the capital in 1872. After that date it began to expand rapidly. Steamships replaced sail and Kingston came into pre-eminence as a seaport, taking full advantage of its site on one of the world's finest natural harbours, and after 1914, of its proximity to the Panama canal. The advent of the railway in the mid-nineteenth century, and its extension to Montego Bay in 1894 and Port Antonio in 1896 consolidated links with the port's hinterland.

Kingston was devastated by an earthquake in 1907, ironically providing an opportunity for extensive rebuilding. The town rapidly expanded its population and geographical area across the alluvial Liguanea Plain around the turn of the century (Clarke, 1975). A network of electric tramways greatly facilitated suburban development and the built-up area soon extended into the lower parts of the neighbouring parish of St Andrew. The combined urban population was 83,000 in 1911, and in 1923, the urban region became known as the Corporate Area. In 1943, its population stood at 237,000.

There was further growth of the urban area in the 1950s and 1960s, and the Corporate Area's population moved from 475,000 in 1972 to 520,000 in 1982. In order to get a more accurate picture of the size of the metropolitan core, however, we need to add the population of Portmore (75,000) and Spanish Town (89,000). Portmore is a rapidly expanding area of dormitory suburbs across Kingston harbour in St Catherine, and the edges of the built-up areas of Kingston and Spanish Town are now only about six miles apart.

The primacy of the Corporate Area is illustrated by the fact in the period 1921 to 1943, its share of the national population moved from less than 10 per cent to around 25 per cent, a value which has remained largely unaltered since. As Jamaica has become more urbanised in recent decades, however, the primacy of Kingston has started to decline in terms of total *urban* population. In 1960, the capital was sixteen times larger than the second ranked town, eleven times greater in 1970, but only six times bigger in 1982. Over the last twenty years, Jamaica's medium sized towns have recorded much faster growth rates than the Corporate Area. For example, between 1970 and 1982, the Corporate Area's population growth rate was only 0.85 per cent per annum, whilst the average rate for Jamaican medium-sized towns was about 4 per cent, and Mandeville and Spanish Town had growth rates as high as 8 per cent and 7 per cent per annum, respectively.

Kingston's emergence as the dominating metropolitan centre owed much to its original population increase by in-migration from rural areas. Migrants were attracted to the capital as it had become the only significant manufacturing and employment centre in the island by mid-century. Hewitt (1974) shows how internal migration into Kingston accelerated through the century. Between 1911 and 1921, migration into Kingston and St Andrew was just under 1,500 per year, but by the late 1940s, the influx was 5,000 per annum. Since then, in-migration accelerated further and was estimated at 11,000 per annum for 1952 to 1959, and 10,000 per annum for the period 1960 to 1970, although more recent data will probably show a decrease in these annual totals for the 1970s.

Artistic and literary impressions add graphically to our comprehension of the context of these migration statistics. For example, the visual romance and trauma of rural migrants' arrival in Kingston and subsequent life in tenement yards and squatter communities is vividly depicted in the film *The Harder They Come*. The film is a fictional representation of the life and death of one of Kingston's earliest legendary gun-men, known by the alias "Rhygin". In the opening scenes, Rhygin arrives in Kingston as a rural migrant on a country bus, and is confronted by a strange urban world in which he must quickly learn how to survive (see also Thelwell, 1980). Another source of rich imagery is a collection of life stories of Jamaican women, produced by a women's theatre collective (Sistren, 1986). The book contains some fascinating and trenchant detail about the move from the country into Kingston, and the adjustment to urban life. The passage below captures some of the excitement and apprehension of the moment of departure to the city, as the narrator awaits the arrival of a country bus, called appropriately (and perhaps with unconscious irony) 'Exodus'.

> "Flamstead Square is where four roads meet. Deh-so dem have di post office, and di variety store dat di Chinee man own. We stand inna di square fi wait pon 'Exodus'. Di whole a we excited. All who go a town come back tell we how town pretty. Kathy go a town and come back wid ring pon her finger tell we seh she married. Me never go a town yet, but me sure me going like it.
> 'Exodus' a come. Di driver draw him brakes and stop. In di dark mum put Tobi ole iron bed pon top a di bus.
> 'Yuh tag everyting?' di conductor ask we. We find we seat and di journey start: 'Exodus' a run". (Sistren, 1986, p.22)

Internal rural-urban migration is only part of a more general population movement from rural areas in Jamaica. It was noted earlier that overseas migration began soon after emancipation. Panama attracted another 45,000 Jamaicans during the period in which the canal was being built, and as the twentieth century progressed, Cuban sugar plantations and the United States became new destinations. International migration took place on a much bigger scale after the Second World War. Between 1945 and the passing of the Commonwealth Immigration Act in 1962, some 200,000 Jamaicans migrated to the United Kingdom. In the period 1960-70 the number of people who migrated overseas was equivalent to 53 per cent of the natural increase in the country's population, and that trend has continued. The

307

annual exodus abroad grew from 17,500 in 1976 to 22,600 in 1984. In 1976, 9,000 of these people went to the United States and 7,300 to Canada, whereas in 1984, the respective values were 19,800 and 2,500. Migration streams can follow step-like patterns, from country district to town and then abroad (Conway, 1983). Jamaican migration follows this route sometimes, but direct movements from rural areas to foreign destinations are also common. Increasingly, return migration seems to be a feature of general, circulatory movements of Caribbean people (Thomas-Hope, 1985). It is a complicating feedback affecting the arena of rural-urban contact since returning migrants are liable to settle in either rural or urban areas. International migration not only affects information flows, however, because remittances from abroad can form an important element of people's income. In 1982, official estimates put the value of remittances at 4.4 per cent of GNP, or nearly 10 per cent of the total value of imports.

The patterns of rural-urban migration became more complex in the second half of the twentieth century because several new regional growth poles appeared, associated with the bauxite and tourist industries. These growth poles also added complexity to overall patterns of rural-urban relations.

Bauxite and alumina operations expanded rapidly after 1952, and throughout the 1960s, Jamaica was the world's leading producer, accounting for over 20 per cent of world production. Export markets are to North America, Western Europe and the U.S.S.R. By 1968, exports of bauxite and alumina accounted for nearly 50 per cent of Jamaica's exports (Jefferson, 1972). In the 1970s the balance of world bauxite production shifted to the vast deposits in Australia and Guinea. Jamaican production began to fall after the introduction of a Bauxite Levy on foreign multinationals in 1974, and the world recession in the 1980s has led to further cutbacks in production.

Bauxite mining is a land extensive operation, a factor which placed the industry in direct competition with agriculture for scarce land resources. In acquiring land for mining, bauxite companies have made a significant impact on the traditional rural economy.

Figure 10.1 shows the past and present distribution of bauxite mining areas and alumina plants. Jamaican bauxite is found in blanket and pocket-like deposits resting on White Limestone formations. Commercial deposits vary in areal size from a few to over seventy acres, and range from five feet to over a hundred feet in depth. There is no overburden; mining operations are open cast and on a large scale. The weight-reduction factor involved in the production of alumina means that alumina plants in Jamaica are also located in the rural mining areas. Bauxite and alumina are exported from specialised ports, built specifically for the task, and which were located away from traditional ports and centres of population.

The industrial plants associated with bauxite mining and alumina production are, in effect, modern, large-scale, capital intensive enclaves, physically located in the traditional rural space economy. As growth poles they do not employ huge industrial labour forces, but the workforce is substantial in terms of local economies. There are no company towns, but Mandeville and to a lesser exent May Pen have experienced growth and prosperity as a result of nearby bauxite operations. Indirect multiplier effects have also benefitted some small towns such as Ewarton and Brown's Town.

Bauxite multinationals began purchasing land for mining operations in the 1940s (Beckford, 1987). Initially, they acquired large properties, particularly cattle pens in the parish of St Ann (Salmon, 1987). In time, and with the entry of other multinationals into Jamaica, land was purchased from peasant farmers.

The process of land acquisition by bauxite companies, and its impact on local rural communities has been documented in detail by Sachak (1987). She argued that the companies bought more land than they actually needed for mining, partly because complete holdings had to be purchased (not all land would have bauxite deposits on it) and partly to strengthen their position *vis a vis* their competitors. In bidding up the price of land and acquiring large tracts, the multinationals became major landowners in Jamaica's rural areas. By the end of the 1970s, they owned nearly 8 per cent of the island's land surface, representing nearly 14.5 per cent of available farm land. The spatial distribution of land ownership was uneven, for example, they owned 48 per cent of the parish of Manchester and 40 per cent of the parish of St Ann. Competition for land and changes in ownership patterns affected farming communities in several ways. In the initial phase, there were numerous tenant small farmers on the large cattle pens, which tended to have absentee owners. Sachak argues that many of the farmers were evicted, became landless, and migrated. In later phases of land acquisition, small farmers were bought out, but bauxite companies developed a variety of settlement schemes to relocate farmers whose land had been purchased. This changed tenure status: small farmers on formerly freehold land became tenants on corporately-owned land. Sachak interprets this as a re-establishment of clientelism in the rural economy, in this case between peasants and foreign multinationals. In effect, bauxite companies became significant agents of land redistribution and tenancy rights in the rural areas.

The dislocation of settled agricultural communities in areas where mining was taking place seems to have affected the flow of rural migrants, both to internal and international destinations, with heavy out-migration from bauxite parishes from the 1950s through to the 1970s (Beckford, 1987). On the other hand, bauxite-alumina centres have themselves attracted migrants, as indicated by the rapid growth of Mandeville from 2,100 in 1943 to 34,500 in 1982, and May Pen's growth from 6,000 to 41,000 over the same time period.

Other impacts of bauxite-alumina corporations on rural communities include shifts in agricultural production patterns in areas affected by mining operations. Originally, much of this land was in typical small farmer land uses, but the rehabilitated mined land is now often used as dairy and beef pasture. The bauxite companies have, themselves, diversified into agriculture, and Alcan has the biggest dairy herd in Jamaica. Finally, rural communities have also been affected by several plant closures, cut-backs in production and cost-cutting exercises during the 1980s, all of which have affected local employment levels in specific areas, but the effects of which have gone largely undocumented.

The second type of regional growth pole which has appeared in Jamaica in the post-war period is associated with the development of tourism. The island has three principal tourist resorts, Montego Bay, Ocho Rios and Negril, and there has been some ribbon development along sections of the north coast. It should be noted that the tourist centres are on the north and west coasts. Kingston is not a major tourist destination (its hotels cater mainly for businessmen), and most of the central and southern parts of the island are relatively unaffected (directly) by tourism.

Jamaica's first tourists visited the island on the returning banana boats at the turn of the century. A few places, such as Port Antonio, acquired reputations as exclusive hideouts for a few wealthy and often famous tourists. The modern tourist period gathered momentum in the 1950s, following technological advances in long range aircraft, and the Cuban Revolution, which lead many North American tourists to seek alternative Caribbean destinations. Tourist arrival statistics fluctuate from year to year, but there has been a trend of steady growth. Numbers passed the half million mark in the early 1970s, and in 1987, Jamaica received over one million tourists for the first time in its history, making the country second only to the Bahamas as a major tourist destination in the English-speaking Caribbean.

Montego Bay was the first resort to rise to prominence, and is still the island's principal tourist area. Over the last thirty years the town has jockeyed with Spanish Town for the distinction of being Jamaica's second largest urban centre. Ocho Rios, on the other hand, was originally a small fishing village. It was transformed into a major tourist centre by the Urban Development Corporation, a Government planning agency. The development of Ocho Rios began at the end of the 1960s and continues today. Its site was deliberately selected at a location where the main road from Kingston to Montego Bay reaches the north coast, and is approximately equidistant from both towns. Negril also came into prominence in the late 1960s, reputedly an idyllic spot "discovered" by hippies. During the 1970s, it rapidly acquired an image and reputation as an 'alternative' resort for young people. The resort, with an impressive four-mile white sand beach is characterised by much haphazard, unplanned, low-rise development and associated informal sector activities (Hudson, 1979).

Much more so than the bauxite centres, the three main tourist resorts have become increasingly significant growth

311

poles. Montego Bay was the first to attract rural migrants (Hewitt, 1974) and in the last 15 years Ocho Rios and Negril have grown largely through in-migration. Formal sector housing has not kept pace with this influx, and the tourist resorts have acquired squatter communities (for a discussion of this phenomenon in Montego Bay see Eyre, 1984).

Direct employment in the tourist industry and indirect multiplier effects have had positive results, but precise figures are difficult to quantify (George, 1987). However, the perception of economic opportunities by actual and potential urban migrants in the Third World is usually highly unrealistic (Barker and Ferguson, 1983). The normal response of rural migrants in developing countries to the lack of formal sector employment opportunities in towns is not to return to the country, but to experiment in income-generating activities. Hence the informal sector has proliferated in all Jamaican tourist resorts. Income-generating activities range from vending food, manufacturing and selling craft items, to drug pushing, prostitution and general tourist harassment. A deep conflict of interests and attitudes has arisen regarding the tourist "product" which is offered for sale between members of the formal and informal sectors. The former constitute key policy makers in the public and private sector whilst the latter group is numerically larger. One manifestation of this polarisation is the appearance of several all-inclusive sanitised tourist packages located in hotel enclaves which are designed to physically isolate tourists from Jamaica's "hustle" economy.

Generally, tourism has tended to disperse wealth and income-generating opportunities away from Kingston, to certain parts of the island's perimeter. Dispersal, however, has retained a distinctive spatial pattern, concentrated at relatively few points. But as tourist centres expand, they acquire their own functional linkages with surrounding rural areas. Commuting patterns are generated, and local farmers have new opportunities to supply food to hotels and restaurants (Belisle, 1984). The appearance of modern tourist growth poles in formerly peripheral areas has encouraged the rapid penetration of materialistic lifestyles and foreign values. This is partly a spin-off from the need to provide and maintain modern tourist infrastructure, and partly the nature of tourism itself, which creates many spatial points of economic and social contact between tourists and both urban and rural Jamaicans.

The rural periphery and environmental degradation

The evolution of Jamaica's rural economy has been reviewed, and in so doing, the inherent spatial duality of a large-scale commercial and a small-scale traditional sector has been noted. The emergence of the Kingston metropolitan core, and smaller regional growth poles associated with the bauxite and tourist industries was also examined. During this gradual transformation of the spatial organisation of Jamaica's economic activities, the small farming sector has become a classic rural periphery in relation to the urban core.

Successive Governments have tried to correct the problems of rural areas and the agricultural sector. For example, sugar production became more mechanised and output increased rapidly after the Second World War, rising to a peak in 1965. But factory closures, declining employment levels and declining acreages in sugar cane have characterised the last twenty years. The fortunes of the banana industry have also fluctuated this century, and it appears that small farmer production has virtually collapsed since Hurricane Allen flattened the banana crop in 1980.

Agricultural policy tends to reflect the prevailing ideology of the political party in power. Thus, in the 1970s, the thrust was towards substantial support for small farmers to try to improve domestic food production and national self reliance (Pollard and Graham, 1985). In the 1980s, attention focussed on the commercial export sector, through a far-reaching programme called AGRO-21, designed to create a hi-tech farming sector based on agribusiness methods (Barker, 1985, 1986). However, successes in agricultural planning have been limited and elusive, particularly with respect to hillside farming. Rural development policy is a labyrinth of ideas, agencies, plans and projects.

In general, the agricultural sector offers few new employment opportunities. The agricultural labour force has only grown from 228,000 in 1943 to 280,000 in 1985, whilst national population increased from 1,237,000 to 2,325,000 over the same period. The majority of Jamaica's rural population is dependent on small farming, but the historically skewed distribution of land holdings has persisted to the present day. There are still 150,000 farmers with less than five acres, representing 80 per cent of the total number of farmers. But these small cultivators farm only 16 per cent of the agricultural land - nearly 65 per cent of agricultural land is on farms over 25 acres, with farms of 100 acres plus accounting for 56 per cent of total agricultural land.

Although the small farm sector is the main source of domestic food supply and grows export crops, hillside farming areas are a true periphery. They are endemic sources of out-migration, demographically characterised by high birth rates and an ageing farm population. They have low income levels, low agricultural productivity and poor levels of social services, educational and medical facilities and other infrastructure.

Conventionally, rural peripheries in developing countries are analysed in terms of their dependent socio-economic and political relationship to the metropolitan core. The physical environment is relegated to a minor role or ignored altogether, although there are current indications of a new cross-fertilisation of ideas (Blaikie, 1985; Blaikie and Brookfield, 1987). An environmental perspective is especially relevant to tropical islands like Jamaica, where the rural periphery is mountainous and soil erosion and land degradation affect the viability of the fragile resource base, and therefore its ability to support an already marginalised rural population.

Further, burgeoning urban areas generate demands not only for fresh farm produce, but they also need a good water supply, timber for the construction industry and an energy supply (which in developing countries translates into charcoal and firewood for many of the urban poor). Clearly, in a situation where these resources are being managed unwisely, for whatever reasons, there is an impact both at source, that is within the rural periphery itself, and in the urban core. The small geographical size of tropical islands in the Caribbean inevitably means that there are environmental and ecological linkages between an island interior, where the majority of the rural population is located, and the coastal zone, where most of the urban population and the best agricultural land is located.

The rest of this section illustrates how environmental deterioration in the rural periphery has an impact outside the rural periphery as well as on the periphery itself. In this way, urban areas can be affected by the environmental status of the rural periphery. Figure 10.2 shows the Kingston urban area and the adjacent rural zone to the immediate east and north east. It forms part of the rugged, ridge and valley topography of the southern flanks of the Blue Mountains, which rise to a maximum height of just over 7,000 feet. The area is noted for the production of high quality Blue Mountain coffee by large and small farmers, and its farmers also supply Kingston's busy markets with agricultural produce. Sections of the region have become functionally part of Kingston's expanding commuter hinterland, and many commuters are part-time farmers.

In the eighteenth century the same region was part of Jamaica's principal coffee plantation area, as mentioned earlier. The steeply dissected topography allowed for only a few small areas of flat land on the top of ridges. They were used as sites for Great Houses and coffee works. The steep slopes bore the brunt of unbridled coffee cultivation for quick profit. Ill-advised agronomic techniques devastated the resource base (McGregor, Barker and Miller, 1985). Later, the post-emancipation rise of peasant farming, based on bush fallowing and the extensive use of fire on steep slopes, caused further accelerated soil erosion and land degradation (Barker and McGregor, 1988).

By the mid-twentieth century, the region was severely depressed, suffering from widespread rural poverty, out-migration and land degradation. After the disruptions caused by a severe hurricane in 1951, a highly innovative rural development strategy unfolded in the 1950s and 1960s. A Land Authority covering the entire Yallahs Drainage Basin was established, and pursued a comprehensive programme of soil conservation, agricultural extension and rural infrastructure provision. Floyd (1970) provided a contemporary review of its progress. Ironically, the impetus for this programme began to fade in the 1970s just as technocrats and planners in other countries began to

315

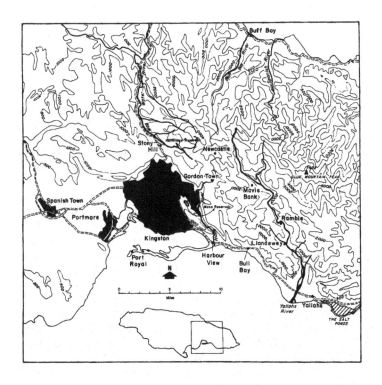

Figure 10.2: The Kingston urban area and its immediate hinterland

recognise and extol the virtues of 'integrated rural development'.

The region remains beset by problems. Small farmers face increasing competition for scarce land from large-scale coffee producers and commercial forestry. McGregor and Barker (1987) document the contemporary environmental and agricultural status of the Fall River basin, a tributary of the Yallahs. Eyre (1987a, 1987b) has written extensively on the problems of deforestation and the use of fire.

The neighbouring Kingston urban area lies in the rainshadow of the Blue Mountains. Rainfall decreases sharply from over 200 inches on the high windward slopes of the Blue Mountains to around 30 inches in Kingston, a straight line distance of less than twenty-five miles. During the last decade or so, Kingston has had recurrent water supply problems, which have meant water lock-offs for extended periods during the dry seasons. The city's current water requirements are about 50 million gallons per day. Half is obained from underground wells and half comes from two reservoirs, Hermitage, built in 1927 and Mona, built in 1957.

Natural fluctuations in rainfall and period of deficit rainfall are mainly responsible for the water shortages. But, increasingly, it is being recognised that environmental degradation in surrounding watershed areas has contributed significantly to the problem. Much of the watershed areas are in agricultural use, and inappropriate land uses and deforestation tend to increase run-off during periods of heavy rainfall. This means that less rainwater infiltrates and percolates into the natural acquifers to recharge them and so less ground water is available to feed rivers on a year round basis. Cunningham (1986) observes that over one hundred rivers and streams have ceased to flow (that is, become ephemeral) in Jamaica during the last fifty years. Thus, less river water is available to replenish the two reservoirs. Also, Hermitage has suffered a 45 per cent reduction in its storage capacity since its construction (Sheng, 1986), caused by silting, another consequence of soil erosion in the surrounding hillsides.

A short-term solution to the urban area's water supply problem was to construct a pipeline from an intake in the Yallahs river to Mona reservoir. In early 1986, the pipeline began to transport 15 million gallons a day into the reservoir. This, together with higher rainfall, seems to have partially alleviated Kingston's water problem, but in turn, has created new problems for small communities of farmers located downstream of the pipeline intake, especially in the very dry Yallahs delta area. Newspaper reports and informal field surveys suggested that during 1984, crops were badly affected in these communities as there was insufficient water in the river to irrigate crops.

Land degradation in the rural periphery can also exacerbate other problems. For example, major floods occurred in Jamaica in June 1986 and October 1987, the former being the more serious. The main causes were slow moving tropical depressions, which are now recognised as posing more serious flood threats to the Caribbean than hurricanes. In the June floods, one station recorded 60 inches of rainfall in two weeks. Whilst this figure is high, it is by no means exceptional in terms of Jamaica's sometimes spectacular rainfall records. Yet the floods were some of the worst in the country's history. The southern plains of Clarendon were particularly badly affected; there was loss of life, a large number of people made homeless, widespread agricultural losses, and damage to buildings, roads, water supply and other infrastructure. Ironically, the June 1986 floods, sounded the death knell for the Spring Plains winter vegetables project. Located on the southern plains, this Israeli farm had been lauded as a showpiece hi-tech farming project (Barker, 1985).

The southern Clarendon plains are drained mainly by the Rio Minho, which has a large catchment area in the upland farming areas of northern Clarendon. Like the Blue Mountains coffee region, the upper watershed areas of the Rio Minho and its tributaries have a long history of land degradation and rural poverty. A Land Authority solution was tried here too, and in the late 1970s, there was an abortive attempt at an integrated rural development project. Although the region is quite productive in agricultural terms, soil erosion and land degradation have not been arrested. Whilst the evidence is not conclusive, it is highly likely that the severity of the floods were made worse by the poor environmental status of large areas of hillsides in the upper reaches of the Rio Minho Basin.

Land degradation in the rural periphery can also cause problems in the offshore marine environment, which may be a potential threat to fishing and tourist industries in the Caribbean. For example, current research at the University of the West Indies indicates that severe damage is taking place to offshore coral reefs outside Kingston Harbour. It seems that during periods of prolonged rainfall and associated flood events, coastal waters become turbid and silty for an extended period, and this affects the optical clarity of sea water. Also, salinity levels decrease temporarily as the floods of fresh water discharged by rivers appear to have a flushing effect on saline coastal waters. The combination of these factors may be a principal cause of mortality in many of the coral reef communities in the area.

Summary

A number of aspects of rural-urban interaction in Jamaica have been examined in this chapter. The approach has been descriptive in order to present material in a straightforward fashion and provide an holistic perspective. The main focus has been on the rural periphery, and on the constraints which are placed on rural-urban interaction on tropical islands like Jamaica because of their small physical size. First, the origins and structure of the rural economy were examined, and then how it became a rural periphery in relation to the dramatic growth of the Kingston metropolitan core. A persistent feature of the periphery has been the exodus of people away from an agrarian way of life, to urban areas both at home and abroad. Latterly, newer regional growth poles have appeared, connected with local developments in the bauxite and tourist industries, and these too are affecting the dynamics of rural-urban interaction. Finally, it was argued that the physical environment needs to be integrated into the functional relationships between urban centres and rural peripheries. One important corollary of the small size of tropical islands like Jamaica, is that environmental degradation can amplify the natural ecological relations between rural and coastal habitats.

Rural-Urban Relations in Jamaica

REFERENCES

Barker, D. (1985) 'New directions for Jamaican agriculture', *Caribbean Geography* 2, 56-64

Barker, D. (1986) 'Recent developments in agricultural planning in Jamaica: The Agro 21 programme', *Paper presented at Caribbena Studies Association meeting*, Caracas, Venezuela

Barker, D. and Ferguson, A.G. (1983) 'A goldmine in the sky faraway: rural-urban images in Kenya', *Area* 15, 185-91

Barker, D. and McGregor, D.F.M. (1988) 'Land degradation in the Yallahs Basin, Jamaica: historical notes and contemporary observations', *Geography* (in press)

Barker, D. and Spence, B. (1988) 'Afro-Caribbean agriculture: a Jamaican maroon community in transition', *Geographical Journal* (in press)

Beckford, G. (1987) 'The social economy of bauxite in the Jamaican man-space', *Social and Economic Studies* 36, 1-56

Belisle, F.J. (1984) 'The significance and structure of hotel food supply in Jamaica', *Caribbean Geography* 1, 219-33

Besson, J. (1984) 'Family land and Caribbean society: towards an ethnography of Afro-Caribbean peasantries', in: Thomas-Hope, E. (ed.) *Perspectives in Caribbean Regional Identity*, University of Liverpool, Centre for Latin American Studies, Monograph Series No.11

Blaikie, P. (1985) *The Political Economy of Soil Erosion*, Longman, London

Blaikie, P. and Brookfield, H. (1987) *Land Degradation and Society*, Methuen, London

Brierley, J.S. (1976) 'Kitchen gardens in the West Indies, with a contemporary study from Grenada', *Journal of Tropical Geography* 43, 30-40

Clarke, C.G. (1975) *Kingston, Jamaica: Urban Development and Social Change, 1692-1962*, University of California Press, Berkeley

Conway, D. (1983) 'The commuter zone as a relocation choice of low income migrants moving in a step-wise pattern to Port of Spain, Trinidad', *Caribbean Geography* 1, 89-105

Cunningham, C.G. (1986) 'Forests and Watershed Management', in: Thompson, D.A., Bretting, P.K. and Humphreys, M. (eds.) *Forests of Jamaica*, Institute of Jamaica, Kingston

Eyre, L.A. (1974) *Geographical Aspects of Population Dynamics in Jamaica*, Florida Atlantic University Press, Boca Raton

Eyre, L.A. (1980) 'The Maroon Wars in Jamaica: a geographical appraisal', *Jamaican Historical Review* 12, 80-102

Eyre, L.A. (1984) 'The internal dynamics of shanty towns in Jamaica', *Caribbean Geography* 1, 256-70

Eyre, L.A. (1987a) 'Fire in the tropical environment', *Jamaican Journal* 20, 10-38

Eyre, L.A. (1987b) 'Jamaica: test case for tropical deforestation?', *Ambio* 6, 336-41

Floyd, B. (1970) 'Agricultural innovation in Jamaica: the Yallahs Valley Land Authority', *Economic Geography* 46, 63-77

George, V. (1987) 'Tourism on Jamaica's north coast: a geographer's view', in: Britton, S. and Clarke, W.C. (eds.) *Ambiguous Alternative: Tourism in Small Developing Countries*, University of the South Pacific, Fiji

Hall, D. (1959) *Free Jamaica 1838-1865: an economic history*, Caribbean Universities Press, Ginn and Company, Aylesbury

Hewitt, L. (1974) 'Internal migration and urban growth', *Recent Population Movements in Jamaica*, World Population Year, C.I.C.R.E.D.

Higman, B.W. (1976) *Slave Population and Economy in Jamaica 1807–1834*, Cambridge University Press

Higman, B.W. (1986) 'Jamaican coffee plantations, 1780–1860: a cartographic analysis', *Caribbean Geography* 2, 73–91

Higman, B.W. (1987) 'The spatial economy of Jamaican sugar plantations: cartographic evidence from the eighteenth and nineteenth centuries', *Journal of Historical Geography* 13, 17–39

Hills, T.L. and Iton, S. (1983) 'A reassessment of the "traditional" in Caribbean small-scale agriculture', *Caribbean Geography* 1, 24–35

Hudson, B.J. (1979) 'The end of paradise: what kind of development for Negril?', *Caribbean Review* 8, 32–33

Jefferson, O. (1972) *The Post-War Economic Development of Jamaica*, I.S.E.R., University of the West Indies, Mona, Kingston

Kopytoff, B. (1973) *The Maroons of Jamaica*, unpublished Ph.D. thesis, University of Pennsylvannia

Lewis, R. (1973) 'Political aspects of Garvey's work in Jamaica, 1929–35', *Jamaica Journal* 7, 30–35

Marshall, W.K. (1972) 'Peasant movements and agrarian problems in the West Indies. Part One: aspects of the development of the peasantry', *Caribbean Quarterly* 18, 30–46

McGregor, D.F.M., Barker, D. and Miller, L. (1985) 'Land resources and development in the Upper Yallahs Valley, Jamaica', *Bedford College Papers in Geography*, No.18

McGregor, D.F.M. and Barker, D. (1987) 'Soil conservation and agricultural development: planning lessons from the Fall River Basin, Yallahs Valley, Jamaica', *Paper presented to the I.B.G. Developing Areas Research Group Conference*, Royal Holloway and Bedford New College, University of London

Mintz, S.W. (1974) *Caribbean Transformations*, Johns Hopkins University Press, Baltimore

Mintz, S.W. (1985) 'From plantations to peasantries in the Caribbean', in: Mintz, S.W. and Price, S. (eds.) *Caribbean Contours*, Johns Hopkins University Press, Baltimore

Pollard, S.K. and Graham, D.H. (1985) 'The performance of the food producing sector in Jamaica, 1962–1969: a policy analysis', *Economic Development and Cultural Change* 33, 731–54

Sachak, N. (1987) 'The impact of land acquisition by bauxite-alumina transnational corporations on the rural economy and society of Jamaica 1942–80', *Social and Economic Studies* 36, 57–92

Salmon, M. (1987) 'Land utilization within Jamaica's bauxite land economy', *Social and Economic Studies* 36, 57–92

Senior, O. (n.d.) 'Slave higglering in Jamaica 1780–1834', *Jamaica Journal* 20, 31–38

Sheng, T.C. (1986) 'Forest land use and conservation', in: Thompson, D.A., Bretting, P.K. and Humphreys, M. (eds.) *Forests of Jamaica*, Institute of Jamaica, Kingston

Simmonds, L. (1987) 'Slave higglering in Jamaica 1780–1834', *Jamaica Journal* 20, 31–38

Sistren (1985) *Lionheart Gal: Life Stories of Jamaican Women* (ed.) Ford-Smith, H., The Women's Press Limited, London

Tekse, K. (1974) *Population and Vital Statistics in Jamaica 1832–1964*, Department of Statistics, Government of Jamaica, Kingston

Thelwell, M. (1980) *The Harder They Come*, Grove Press Inc., New York

Thomas-Hope, E. (1985) 'Return migration and its implications for Caribbean Development', in: Pastor, R. (ed.) *Migration and Development in the Caribbean*, Holmes and Meier, New York

11

URBAN-RURAL INTERACTION, SPATIAL POLARISATION AND DEVELOPMENT PLANNING

ROBERT B. POTTER

The types of interaction occurring between urban and rural areas in developing countries, and their relative magnitudes, are predicated upon the degree of socio-economic disparity which exists between these ostensibly contrasting portions of the national space. Expressed in the simplest terms, the greater the spatial disparities, the larger the flows and interactions which serve to promote and maintain them. Rather than stressing the salience of these urban-rural territorial differences, however, the chapters included in this book serve to exemplify the universality and essential unity of the social, economic, political and administrative processes which have etched-out these spatial contrasts.

It is in this connection that discussion of the empirical character of rural-urban interactions, linkages and flows in Third World countries merges with the theoretical considerations which surround the issues of development planning, in both its territorial and sectoral forms. Thus, urban-rural interaction involves the study of relative degrees of urban bias, the degree to which 'top-down' and 'centre-out' strategies of change have been pursued, either implicitly or explicitly, as opposed to the reverse 'bottom-up' and 'periphery-in' path to national development planning. In short, efforts to maintain the existing settlement system, or indeed to change it, are inextricably bound-up with societal goals in general. In particular, these wider aims involve fundamental ideological and political choices concerning the economic efficacy and social morality of personal and territorial welfare differences, and the overarching concepts of social justice and equity. It is in this connection that the adoption of structuralist and political economy approaches to the study of development issues, stressing the role of the spatial division of labour, is starting to move work toward a more specific consideration of the character of urban-rural relations and interactions.

Thus, a growing body of opinion now argues that in the past, development planning practice in capitalist countries has been insufficiently concerned with promoting the types of urban-rural links and interactions which generate genuine patterns of development and social change. Rather, urban to urban, or at best, large urban to smaller urban growth impulses have been promoted and sustained. Rural areas have been regarded as true peripheries in all respects, that is areas which have been left behind, rather than held back by the accumulation of capital elsewhere. This explains the focus on national urban development strategies in Third World countries which stress principles of what may best be described as 'concentrated deconcentration', and not true decentralisation and deconcentration *per se*. The studies included in this book have alluded to many instances of this type. Thus, in North Africa and the Caribbean, as indeed elsewhere, it appears that many new economic developments in the fields of tourism, manufacturing and mining, whilst being held up as ostensibly spreading economic development, in reality, have largely served to further concentrate decision-making and capital, both socially and/or spatially. In other contexts too, the development of new techniques has apparently served to increase capital accumulation and has, if anything, promoted the net outflow of further resources from rural regions and thereby, greater urban competition for rural resources. This applies to the use of fuelwood in the African setting, and to the attraction of migrant labour, even on an international basis. The stimulation of rural agricultural products at low procurement prices to serve urban markets is yet another instance of the operation of the same kinds of processes. All these situations witness the continued operation of a centre-oriented pattern of capitalistic growth and the evolution of more efficient means of centralising and controlling the expropriation of surplus product from labour.

324

Consumption, Production and National Space in Developing Countries

Ideas of spatial dependency theory provide an interesting framework for the examination of rural-urban interaction in developing countries (see Figure 9.2), although subsequently in any such formulation, every effort must be made to include national influences such as the development of local social elites (see Forbes, 1984; Potter and Binns, 1988; Vendovato, 1986; Sandbrook, 1985). Viewed in territorial terms, a process model is revealed whereby capital is extracted from the rural periphery and successively moved toward the urban centre and thence to the metropole. In theory, the corollory of this short-term spatial polarisation is the eventual diffusion of growth to the rural periphery. This it is posited will occur after some point of spontaneous polarisation reversal has been reached (Richardson, 1980; Alonso, 1980; Friedmann, 1966).

But for many commentators, post-war experience has been enough to suggest that these long-term 'generative', 'trickling down' or 'spread' effects do not appear to be stimulated without the positive and direct intervention of the state. Thus, 'parasitic' flows, 'backwash' and 'polarisation' seem all too often to have been the order of the day. Further, the introduction of new micro-electronics based industrial opportunities seems to be doing little or nothing to change the situation. The global mobility of multinational capital appears to be leading to further inequalities rather than their eradication or even reduction. Thus, although generally in the past optimistic about the continuing role of large cities in generating and spreading growth in developing countries, the regional economist Harry Richardson (see Clapp and Richardson, 1984) has recently argued that the micro-electronics revolution, far from reducing the tyranny of distance and affording the opportunity for growth without propinquity, is likely to enhance further the economic importance of core regions. This mainly reflects the situation whereby unit production costs are always likely to favour the core urban areas where infrastructure has already been installed in Third World countries.

This brings us to a very important matter, that of the precise nature of the things that are diffused from, and/or concentrated in cities of the developing world. It is increasingly tempting to suggest that in these socio-economic circumstances, cities are primarily serving to centralise, rather than diffuse, capital, decision-making, production, incomes and welfare. This in itself represents a far cry from the early and optimistic assertion that innovation paths within an urban system are almost perfectly hierarchical, so that the national diffusion of a new technique or idea would merely be a matter of time (Hudson, 1969; Pedersen, 1970; Berry, 1972; Robson, 1973). It was Pred (1973, 1977) who, on the basis of developed world evidence, suggested strongly that in the modern world, growth-inducing innovations are likely to be bounced around between large places, rather than spread by processes of either spatial or hierarchical diffusion.

On the other hand, cities in the Third World may increasingly be seen as the centres from which the norms of global capitalist consumption are diffused to surrounding populations. This strongly dependency-related thesis has been articulated most cogently by Armstrong and McGee (1985) in their study of what they describe as *Theatres of Accumulation*. Thus, they maintain that "Cities are, stated simply, the crucial elements in accumulation at all levels, ... providing .... the *locus operandi* for transnationals, local oligopoly capital and the modernizing national state"; although "On the other hand, cities also play the role of diffusers of the lifestyles, customs, tastes, fashions and consumer habits of modern industrial society. These two processes of centralization and diffusion are, of course, not contradictory for the expansion of markets is a necessary part of growth in the capitalist system" (Armstrong and McGee, 1985, p.41). Thus, the most important role of Third World urban systems is that of "bases for centralized capital accumulation by transnational firms and local businesses" (Armstrong and McGee, 1985, p.44).

It is precisely this line of argument that has led to the call for a total reappraisal of urban-rural interactions, spatial polarisation and indeed, overall paths to development in Third World countries. The basic argument has been outlined effectively by Friedmann and Weaver (1979), where they stress that in such contexts, development planning needs to be based on territorial objectives and not purely functional imperatives. In other words, following the prescriptions of dependency theory, namely that generally, the more nations have become involved with the capitalist system, the more underdeveloped they have become, the clarion call is to develop from the rural areas outward on an essentially indigenous basis. It is posited that only in this manner can the socially regressive forces of capital accumulation be ameliorated, with their implicit connotations of uneven spatial development and the strongly parasitic rural-urban interactions and flows which maintain them.

Thus, a catch phrase is the provision of *basic needs* at a regional level following *agropolitan principles* of growth. This involves the promotion of *selective regional closure*, whereby external trade links are cut and rural regions are thereby sheltered from the diffusion of western lifestyles and consumption norms. This happens until basic needs have been provided for, and the rural areas manage to catch-up. The crucial dilemma which is faced by poor dependent countries was stressed some time ago by Lasuen (1973). In theory at least, such nations can import new techniques and technologies in an unbridled manner and hope that spontaneous processes of spatial-cum-hierarchical diffusion and polarisation reversal will eventually afford their peripheral regions the chance of catching-up. On the other hand, they can wait for the previous set of innovations to spread fully, before the next round is introduced. But given the argument rehearsed previously, that unit costs are always likely to be appreciably lower where infrastructure already exists, it is hard to imagine that even governments in countries wishing genuinely to promote more equitable patterns of growth and development have much effective choice in the matter if they do opt for the wholesale introduction of new productive techniques. Further, the extremely hypothetical nature of this entire debate is revealed when the scale and rate of contemporary entrepreneurial and domestic technological change is considered, along with the influence of multinational companies and international capital. The time lag for spatial spread effects to occur is likely to be inordinately long.

A vital point in this regard is that it is the wider development path adopted by nations that is crucial. The basic distinction is specifically between the top-down and quintessentially *laissez-faire*, capitalist model of development planning on the one hand, and the bottom-up, interventionist and basically socialist formulation on the other. Although highly polemical, some might summarise these as the 'spread wealth unequally' and the 'share poverty equally' theses of development planning respectively. Of course, the humanitarian aim must ultimately be to 'spread welfare and opportunity equally'.

The bottom-up ideology suggests that when the time is right to diversify the economy and non-agricultural activities are thereby eventually introduced, urban locations will no longer be mandatory. Thus, on the one hand the thesis posits that cities can quite easily be based on agricultural foundations. On the other, it is suggested that "large cities will lose their present overwhelming advantage" (Friedmann and Weaver, 1979, p.200).

Walter Stöhr (1981) has prodvided an informative account concerning the nature of the 'development from below' strategy. In particular, his essay stresses 'that there is no categorical receipe for such approaches, as is the case with those 'from above'. Development from below needs to be closely and sensitively dovetailed with the specific socio-cultural, historical and institutional conditions prevailing in each and every country. But Stöhr does stress that under such a strategy, development should be based on territorial units and should endeavour to mobilise both their natural and human resources to the fullest extent possible. Perhaps most usefully, Stöhr lists what he sees as the eleven essential components of such change, and the connotations these carry for the relations and interactions existing between rural and urban areas are clearly evident:-

1. broad access to land (involving land reform);
2. a territorially organised structure for equitable communal decision-making;
3. granting greater self-determination to rural areas;
4. selecting regionally appropriate technology;
5. giving priority to projects which serve the satisfaction of basic needs (food, shelter, maximising the use of the informal sector);
6. the introduction of national pricing policies (offering more favourable terms of trade to the periphery);

7. external resources to be employed only where rural-peripheral ones are inadequate;
8. the development of productive activities exceeding regional demands;
9. restructuring urban and transport systems to include all internal regions;
10. improving rural-to-urban and village transport and communications;
11. encouraging egalitarian societal structures and collective consciousness.

In the period since the Second World War, the countries which have genuinely followed the bottom-up model of development planning, in an effort to reduce the effects of large metropolitan cities and their ties with capital accumulation and western patterns of consumption, have generally been socialist states. Although previously neglected, their experiences have been summarised in a number of recently published works (Drakakis-Smith, Doherty and Thrift, 1987; Forbes and Thrift, 1987).

The transition to socialism frequently involves avowedly anti-urban policies. Illustrative of this trend are programmes of de-urbanisation involving the enforced movement of urbanites to the countryside, as in the drastic instance of Cambodia (Kampuchea). In other cases, city size and/or city growth rates may be reduced, as the experiences of countries such as China, Vietnam and Zimbabwe bear witness (Forbes and Thrift, 1987). This is, of course, principally because large urban places are seen as centres of consumption. But socialist states are not merely worried about the relative spatial concentration of production and consumption. They are concerned just as much about the fact that large cities are the spatial bases of the bourgeoisie, along with what are seen as decadent and corrupting ways of life (Mingione, 1981). Thus, as in the case of Mao's China, favouring agriculture and fostering rural identity is typical of the socialist development strategy. The ultimate aim is a new territorial order, specifically one which is no longer based on a clear and dichotomous division between rural and urban components of the national space. Indeed, many recent writers have been at pains to stress that both Marx and Engles saw rural-urban relations and interactions as

329

primarily exploitive, and cherished the ultimate objective of fusing town and country in such a manner that social equality would be matched by spatial equality. Many of these issues concerning urban-rural relations and the wider socio-political remit of development planning are exemplified by the post-revolutionary experiences of Cuba. The overall aim in Cuba has undoubtedly been to urbanise the countryside and ruralise the towns (Gugler, 1980). In a recent account, Susman (1987) has stressed how since 1959 spatial equality has been an intrinsic part of Cuban development planning, with programmes specifically designed to overcome rural-urban differences at the local level being prominent. In this respect, the new policies were basically aimed at transforming social and spatial patterns which were the outcome of four centuries of Spanish rule and 60 years of domination by the United States. The main elements involved in the change were an emphasis on production rather than consumption, and efforts to integrate town and country. Thus, "Unlike many other countries limiting their goals to increasing *consumption per capita* via the distribution of goods and services, Cuban objectives include increasing equality in participation in decisions concerning *production* activities over the population and country as a whole" (Susman, 1987, p.251). Thereby, it is argued that ".... many of the changes in Cuban spatial organization may be linked to the goal of overcoming urban-rural disparities in both consumption and production spheres" (Susman, 1987, p.253).

These goals have been pursued by means of a wide variety of programmes. Rather than endeavouring to reduce urbanisation *per se*, these were specifically aimed at increasing national integration. But having said that, in order to achieve these goals, strong efforts were made to reduce Havana's urban primacy by disinvesting in the capital city. Thus, as from 1963, little in the way of new public works or investment were centred on Havana, and the provision of new houses and jobs there all but ceased. At the same time, other policies sought to emphasise the growth of new provincial towns in the 20,000-200,000 size range, in a direct effort to counterbalance the capital, but without establishing very large urban centres in so doing (see Potter, 1985, p.130-31). Also, village settlements were reorganised into so called "rural towns" in an effort to upgrade rural living conditions (Hall, 1981). The establishment of local central place hierarchies based on provincial capitals also aimed to increase the integration existing between rural and urban landscapes and ways of life.

But most saliently, every effort was made to redistribute social surplus product to rural areas. The provision of services in rural areas was seen as a major priority, particularly with regard to primary health care facilities. This was backed by the provision of secondary and tertiary health care units in towns and cities throughout the country. A similar rural emphasis was true of the provision of schools. In 1971/72, when the programme got underway, only seven out of 478, or 1.5 per cent of the nation's secondary schools were situated in rural areas. But quite remarkably, by 1979/80, 533 of 1,318, or 40.4 per cent of such schools were to be found in the countryside (Susman, 1987). More specifically, education was deliberately used "to bring urban and rural populations together as a means of overcoming attitudes of domination and subjugation" (Susman, 1987, p.270). This included a programme whereby all students were at some stage integrated into the productive agricultural process. These sections of the urban population were temporarily shifted to rural areas in a direct bid to eradicate what were seen to be their essentially elitist attitudes and values.

Conclusion

The aim of promoting greater equality between urban and rural areas, so central to the Cuban experience briefly reviewed above is, of course, pivotal to the whole issue of urban-rural interaction in developing countries and to the overall theme addressed in this book.

The need for an integrated approach to the analysis of what may superficially appear to represent quite separate and contrasting portions of national space is a major conclusion. Such an integrated spatial approach is required both in relation to the directions which may be taken by future academic research, and with regard to the formulation of appropriate and effective development and planning policies in developing countries. It is not the nature of the settlement system of a country that is the source of the planning problems it faces - ones which are to be ameliorated by specific urban solutions such as growth poles, new towns, pronouncements on optimal city size, or the establishment of a new provincial or capital city. Rather, these problems merely reflect the existence of deeper socio-economic difficulties and inequalities. Realistically, more fundamental socio-political issues are of crucial significance, not least, the mode of production and the development path that have been adopted in particular countries.

This line of argument also serves to emphasise once more the salience of the territorial dimension to development planning in Third World countries. In turn, this implies that geographers have an important role to play in the study of appropriate forms of social development and change in developing countries. Geographers more than most are well placed to analyse the closely interlocking spatial elements of the development equation - urban and rural areas - in a manner which recognises their fundamentally integrated and holistic nature.

Urban-Rural Interaction and Planning

REFERENCES

Alonso, W. (1980) 'Five bell shapes in development', *Papers of the Regional Science Association* 45, 5-16

Armstrong, W. and McGee, T.G. (1985) *Theatres of Accumulation: Studies in Asian and Latin American Urbanization*, Methuen, London

Berry, B.J.L. (1972) 'Hierarchical diffusion: the basis of development filtering and spread in a system of growth centres', in: Hanson, N.M. (ed.) *Growth Centres in Regional Economic Development*, The Free Press, New York

Clapp, J.M. and Richardson, H.W. (1984) 'Technological change in information-processing industries and regional income differentials in developing countries', *International Regional Science Review* 9, 241-56

Drakakis-Smith, D.W., Doherty, J. and Thrift, N. (1987) 'Introduction: what is a socialist developing country?', *Geography* 72, 333-35

Forbes, D.K. (1984) *The Geography of Underdevelopment: A Critical Study*, Croom Helm, London

Forbes, D.K. and Thrift, N. (eds.) (1987) *The Socialist Third World: Urban Development and Territorial Planning*, Basil Blackwell, Oxford

Friedmann, J. (1966) *Regional Development Policy: A Case study of Venezuela*, M.I.T. Press, Cambridge, Massachusetts

Friedmann, J. and Weaver, C. (1979) *Territory and Function: the Evolution of Regional Planning*, Arnold, London

Gugler, J. (1980) 'A minimum of urbanism and a maximum of ruralism: the Cuban experience', *International Journal of Urban and Regional Research* 4, 516-35

Hall, D.R. (1981) 'Town and country planning in Cuba', *Town and Country Planning* 50, 81-83

Hudson, J.C. (1969) 'Diffusion in a central place system', *Geographical Analysis* 1, 45-58

Lasuen, J.R. (1973) 'Urbanisation and development - the temporal interaction between geographical and sectoral clusters', *Urban Studies* 10, 163-88

Mingione, E. (1981) *Social Conflict and the City*, Blackwell, Oxford

Pedersen, P.O. (1970) 'Innovation diffusion within and between national urban systems', *Geographical Analysis* 2, 203-54

Potter, R.B. (1985) *Urbanisation and Planning in the Third World: Spatial Perceptions and Public Participation*, Croom Helm, London and St Martin's Press, New York

Potter, R.B. and Binns, J.A. (1988) 'Power, politics and society', in: Pacione, M. (ed.) *The Geography of the Third World: Progress and Prospects*, Routledge, London, 271-310

Pred, A.R. (1973) 'The growth and development of systems of cities in advanced economies', in: Pred, A.R. and Tornquist, G., *Systems of Cities and Information Flows: Two Essays*, Lund Studies in Geography, Series B, 38, 9-82

Pred, A. (1977) *City-Systems in Advanced Economies*, Hutchinson, London

Richardson, H.W. (1980) 'Polarisation reversal in developing countries', *Papers of the Regional Science Association* 45, 67-85

Robson, B.T. (1973) *Urban Growth: An Approach*, Methuen, London

Sandbrook, R. (1985) *The Politics of Africa's Economic Stagnation*, Cambridge University Press

Stöhr, W.B. (1981) 'Development from below: the bottom up end periphery-inward development prardigm', in: Stöhr, W.B. and Taylor, D.R.F. (eds.) *Development from Above or Below?*, John Wiley, Chichester

Susman, P. (1987) 'Spatial equality and socialist transformation in Cuba', in: Forbes, D. and Thrift, N. (eds.) *The Socialist Third World: Urban Development and Territorial Planning*, Blackwell, Oxford

Vendovato, C. (1986) *Politics, Foreign Trade and Economic Development: A Study of the Dominican Republic*, Croom Helm, London

INDEX

A

Absentee land ownership 80-81
Accumulation
  capital 250, 266, 324, 326-327
Administrative provision 12
Africa 11, 14, 17, 26, 28, 29, 235,
  243, 252
  Maghreb 28, 68-107, 324
  southern 28, 141-168
  sub-Saharan 222-223, 243, 247
  Tropical 33-67, 109-140
  west 29, 23, 38, 44, 47, 50-51, 56,
  59, 61-63, 133, 204, 214-217,
  226
Agrarian reform 28, 76-84, 88, 101-
  104, 147, 160-163
Agriculture 16, 18-19, 24, 29, 50,
  56, 58, 62, 72, 74-100, 103-
  104, 118, 121-122, 124, 129-
  130, 132-135, 144, 147, 150-
  154, 157, 160, 169, 171, 174-
  182, 188, 192-201, 204, 214,
  217-221, 238, 241-244, 247,
  249-251, 253, 270, 295, 310,
  313
  capitalist 217-221
  pastoral 80-81, 174-182, 197
  plantation 263, 297-304, 315
Agroforestry 62
Aid 159
Algeria 68-79, 80-85, 91, 93-99,
  101-105
Angola 46, 50
Appropriate technology 22, 62
Arabian peninsula 20, 29, 127, 132,
  135, 169-203
Asia 28-29, 235, 247, 251
  south 20, 29, 204, 210, 217-227,
  245, 248-252
Autogestion 76

B

Backwash effects 14, 98, 272, 325
Bahrain 172-178, 183-184, 186-187,
  189-191, 198-199
Bananas 303-304, 313
Banking 124, 152, 188, 219
Bantustans 141, 146
Barbados 29, 257-293
Basic needs 15, 24, 34, 289, 327
Bauxite mining 302, 308-310, 313,
  319

Beckford, G. 258, 263-265, 270, 289
Bedouin 175-177, 181-182, 192
Biomass fuel 38, 47-48
Botswana 148
Bottom-up development 13, 15, 18,
  22, 24-25, 100-102, 137, 289,
  323, 327-328
Britain 275, 297

C

Capital 197, 199, 207, 237, 247, 251,
  266
  accumulation 250, 266, 324, 326-
  327
  flows 12, 48, 91, 258
  markets 187
  multinational 97, 272, 309-310,
  325
Capitalism 18-21, 143, 151, 213,
  235, 239-240, 248, 254, 266-
  267, 326
Capitalist farming 217-221
Capitalist mode of production 12,
  151, 156-157, 213, 324, 328
Caribbean 20, 28-30, 257-322, 324
Cash 206, 209, 214
  flows 26, 48
  crops 86, 129, 151, 206
Central-place hierarchy 239, 331
Centripetal forces 268
Chambers, R. 15, 34
Charcoal 40-44, 50-52, 54, 58, 60
Children
  nutrition of 224
China 243, 251, 329
Cities
  generative 13, 15, 19, 213
  parasitic 13, 15, 18-19, 118, 213
  primate 20, 94, 97, 102, 105, 112,
  127, 130, 136, 260-261, 294,
  305-306
  secondary 18-21, 101, 105
City size hierarchy 18-21, 100, 116,
  137, 326
Class 24-26, 58, 157, 221, 226, 248,
  250, 254
Coastal development 72, 94, 174,
  281, 296, 301, 310-315
Colonial legacy 29-30, 68-69, 76,
  79-80, 84, 91, 111, 136, 143,
  148, 155-156, 158, 175, 177,
  235, 248, 257, 262-265, 270,
  295, 297-304
Commerce 209, 258
Commercialisation 204-232
Commodity
  flows 211-212, 227
  production 207, 219
Communications 101, 111, 236

Index

Index

Index

Index

Multinational companies 97, 272, 309-310, 325

N

Namibia 141
Neo-classical economics 29, 100, 233, 239-244, 247, 253
Neo-colonialism 270
Niger 29, 204, 214-217, 226
Nigeria 29, 34, 38, 44, 47, 50-51, 56, 59, 61-63, 204, 214-217, 226
Nomadic pastoralism 98, 125-126, 133, 135, 174-182, 190, 192, 197, 200
Nomads 175-176, 177, 181-182
Nutrition 29, 42, 204-205, 221-226
children 224

O

Oil 54-56, 71, 98, 126, 172, 177, 182-190, 192, 200
prices 33, 42, 54, 56, 61, 95, 183, 186, 193
refining 184, 186-187
Oman 172-173, 178, 183-184, 189-192, 198
Organization of the Petroleum Exporting Countries 182-183
Outsiders 34

P

Parasitic cities 13, 15, 18-19, 118, 213
Pastoral farming 80-81, 174-182, 197
Pasture 80-81, 174-182
Pearling 174, 176
Peasant 267
Peasant households 204, 221-225, 242
Peasant production 144, 146-147, 150-154, 156-158
Peasant society 204-232, 264
Perception 265, 279, 289
Peripheries 266, 294-295, 314, 318-319, 324
Petrochemicals 71, 182-189
Petty commodity production 211-212
Planning 193, 245, 248-249, 257, 313
Plantation economy 258, 263, 267, 269, 275, 295, 297-304
Politics 17, 24, 93, 98, 104, 128, 144, 163, 212, 234, 236, 244-245, 250

Pollution 47
Population 152, 193, 294-295, 304, 306
rural 35, 269, 304, 314
urban 18, 35-38, 48, 51, 69, 112, 116-121, 124-127, 169, 171-174, 193, 198, 259-261, 305-306
Portugal 148, 150
Poverty 11-12, 16, 19, 21, 28, 54, 137-138, 163, 240, 258, 265, 269
rural 34, 100
Precapitalist modes of production 205, 208
Prices 151, 206-207, 216
food 16-17, 50, 74-75, 103, 196, 238, 241, 243
fuel 39-45, 50, 54, 60-61, 64
oil 33, 42, 54, 56, 61, 95, 183, 186, 193
Pricing policies 103, 206, 241
Primate cities 20, 94, 97, 102, 105, 112, 127, 130, 136, 260-261, 294, 305-306
Proletarianisation 157, 161-162
Proletarians 209
Prostitution 273, 312

Q

Qatar 172-173, 176, 178, 183-187, 189-191, 199

R

Rank-size rule 20
Refugees 110
Regional development 21, 33, 68, 72, 75, 91, 93-100, 149
Regional inequalities 19, 129
Religion 174, 300
Islam 127, 174, 181, 224
Remittances 134, 144, 147, 155, 158, 190, 308
Rent 205
Resource transfer 12, 38, 91
Resources 16, 22, 24-25, 29, 38, 48, 118, 134, 176, 205, 209, 294, 314
flows 38, 91
rural 75, 84-91
water 75, 84-91, 118, 175, 178, 181, 193, 198-200
Retail facilities 258
Retailing 267, 277, 279, 281-283
Return migration 147, 150, 153, 155, 159, 163

Index

Index

Urban primacy 20, 94, 97, 102, 105, 112, 127, 130, 136, 260-261, 294, 305-306
Urban water demands 88, 91, 317
Urbanisation 35-38, 60, 62, 69-70, 95, 101-105, 109-140, 161, 171-174, 177, 189, 196, 198, 200, 244, 259-269
balanced 20
Usury 205 210

V

Vance, J.E. 261-262
Vernacular architecture 269
Vineyards 74

W

Wages 56, 146, 148, 152, 154-155, 157, 158
War 125-126, 129, 136, 150, 159, 163
Water 34, 199, 217
desalination 199
resources 75, 84-91, 118, 175, 178, 181, 193, 198-200, 317-318
shareholding 181
Wine 74
Women 160-161, 212, 307
labour 223-225
liberation 159-160
nutrition of 222-225
Woodfuel 33, 38-45, 48-52, 54, 56, 58-60, 62-64, 324

Y

Yemen Arab Republic 172-174, 176, 178, 183-184, 189-191
Yemen, People's Democratic Republic of 183-184, 186, 189-190

Z

Zambia 143, 148, 153
Zimbabwe 28, 47, 61, 141, 147-148, 153-154, 158, 160-164, 329

Printed in the United States
by Baker & Taylor Publisher Services